203
Advances in Polymer Science

Editorial Board:
A. Abe · A.-C. Albertsson · R. Duncan · K. Dušek · W. H. de Jeu
J.-F. Joanny · H.-H. Kausch · S. Kobayashi · K.-S. Lee · L. Leibler
T. E. Long · I. Manners · M. Möller · O. Nuyken · E. M. Terentjev
B. Voit · G. Wegner · U. Wiesner

A·NAL

QD
281
P6
F66
v. 203
CHEM

Advances in Polymer Science
Recently Published and Forthcoming Volumes

Chemistry Library

QD
281
PG
F66
V. 203
CHEM

Polymers for Regenerative Medicine

Volume Editor: Carsten Werner

With contributions by

J. H. Elisseeff · C. Fischbach · T. Freier · A. J. García · D. J. Mooney
T. Pompe · K. Salchert · S. Varghese · C. Werner · S. Zhang · X. Zhao

 Springer

The series *Advances in Polymer Science* presents critical reviews of the present and future trends in polymer and biopolymer science including chemistry, physical chemistry, physics and material science. It is adressed to all scientists at universities and in industry who wish to keep abreast of advances in the topics covered.
As a rule, contributions are specially commissioned. The editors and publishers will, however, always be pleased to receive suggestions and supplementary information. Papers are accepted for *Advances in Polymer Science* in English.
In references *Advances in Polymer Science* is abbreviated *Adv Polym Sci* and is cited as a journal.

Springer WWW home page: springer.com
Visit the APS content at springerlink.com

Library of Congress Control Number: 2006927158

ISSN 0065-3195
ISBN-10 3-540-33353-3 Springer Berlin Heidelberg New York
ISBN-13 978-3-540-33353-1 Springer Berlin Heidelberg New York
DOI 10.1007/11604228

This work is subject to copyright. All rights are reserved, whether the whole or part of the material is concerned, specifically the rights of translation, reprinting, reuse of illustrations, recitation, broadcasting, reproduction on microfilm or in any other way, and storage in data banks. Duplication of this publication or parts thereof is permitted only under the provisions of the German Copyright Law of September 9, 1965, in its current version, and permission for use must always be obtained from Springer. Violations are liable for prosecution under the German Copyright Law.

Springer is a part of Springer Science+Business Media

springer.com

© Springer-Verlag Berlin Heidelberg 2006
Printed in Germany

The use of registered names, trademarks, etc. in this publication does not imply, even in the absence of a specific statement, that such names are exempt from the relevant protective laws and regulations and therefore free for general use.

Cover design: *Design & Production* GmbH, Heidelberg
Typesetting and Production: LE-TEX Jelonek, Schmidt & Vöckler GbR, Leipzig

Printed on acid-free paper   02/3100 YL – 5 4 3 2 1 0

Volume Editor

Prof. Dr. Carsten Werner
Leibniz Institute of Polymer
Research Dresden
Max Bergmann Center of Biomaterials
Hohe Str. 6
01069 Dresden, Germany
werner@ipfdd.de

Editorial Board

Prof. Akihiro Abe
Department of Industrial Chemistry
Tokyo Institute of Polytechnics
1583 Iiyama, Atsugi-shi 243-02, Japan
aabe@chem.t-kougei.ac.jp

Prof. A.-C. Albertsson
Department of Polymer Technology
The Royal Institute of Technology
10044 Stockholm, Sweden
aila@polymer.kth.se

Prof. Ruth Duncan
Welsh School of Pharmacy
Cardiff University
Redwood Building
King Edward VII Avenue
Cardiff CF 10 3XF, UK
DuncanR@cf.ac.uk

Prof. Karel Dušek
Institute of Macromolecular Chemistry,
Czech
Academy of Sciences of the Czech Republic
Heyrovský Sq. 2
16206 Prague 6, Czech Republic
dusek@imc.cas.cz

Prof. W. H. de Jeu
FOM-Institute AMOLF
Kruislaan 407
1098 SJ Amsterdam, The Netherlands
dejeu@amolf.nl
and Dutch Polymer Institute
Eindhoven University of Technology
PO Box 513
5600 MB Eindhoven, The Netherlands

Prof. Jean-François Joanny
Physicochimie Curie
Institut Curie section recherche
26 rue d'Ulm
75248 Paris cedex 05, France
jean-francois.joanny@curie.fr

Prof. Hans-Henning Kausch
Ecole Polytechnique Fédérale de Lausanne
Science de Base
Station 6
1015 Lausanne, Switzerland
kausch.cully@bluewin.ch

Prof. Shiro Kobayashi
R & D Center for Bio-based Materials
Kyoto Institute of Technology
Matsugasaki, Sakyo-ku
Kyoto 606-8585, Japan
kobayash@kit.ac.jp

Prof. Kwang-Sup Lee

Department of Polymer Science &
Engineering
Hannam University
133 Ojung-Dong
Daejeon 306-791, Korea
kslee@hannam.ac.kr

Prof. L. Leibler

Matière Molle et Chimie
Ecole Supérieure de Physique
et Chimie Industrielles (ESPCI)
10 rue Vauquelin
75231 Paris Cedex 05, France
ludwik.leibler@espci.fr

Prof. Timothy E. Long

Department of Chemistry
and Research Institute
Virginia Tech
2110 Hahn Hall (0344)
Blacksburg, VA 24061, USA
telong@vt.edu

Prof. Ian Manners

School of Chemistry
University of Bristol
Cantock's Close
BS8 1TS Bristol, UK
ian.manners@bristol.ac.uk

Prof. Martin Möller

Deutsches Wollforschungsinstitut
an der RWTH Aachen e.V.
Pauwelsstraße 8
52056 Aachen, Germany
moeller@dwi.rwth-aachen.de

Prof. Oskar Nuyken

Lehrstuhl für Makromolekulare Stoffe
TU München
Lichtenbergstr. 4
85747 Garching, Germany
oskar.nuyken@ch.tum.de

Prof. E. M. Terentjev

Cavendish Laboratory
Madingley Road
Cambridge CB 3 OHE, UK
emt1000@cam.ac.uk

Prof. Brigitte Voit

Institut für Polymerforschung Dresden
Hohe Straße 6
01069 Dresden, Germany
voit@ipfdd.de

Prof. Gerhard Wegner

Max-Planck-Institut
für Polymerforschung
Ackermannweg 10
Postfach 3148
55128 Mainz, Germany
wegner@mpip-mainz.mpg.de

Prof. Ulrich Wiesner

Materials Science & Engineering
Cornell University
329 Bard Hall
Ithaca, NY 14853, USA
ubw1@cornell.edu

Advances in Polymer Science
Also Available Electronically

For all customers who have a standing order to Advances in Polymer Science, we offer the electronic version via SpringerLink free of charge. Please contact your librarian who can receive a password or free access to the full articles by registering at:

springerlink.com

If you do not have a subscription, you can still view the tables of contents of the volumes and the abstract of each article by going to the SpringerLink Homepage, clicking on "Browse by Online Libraries", then "Chemical Sciences", and finally choose Advances in Polymer Science.

You will find information about the

– Editorial Board
– Aims and Scope
– Instructions for Authors
– Sample Contribution

at springer.com using the search function.

Preface

New insights in cell and developmental biology are expected to pave the way for new therapies for a number of severe diseases by exploiting mechanisms to reactivate the cellular potential of regeneration. However, the utilization of molecular switches controlling cellular fate decisions for new therapies very often depends on the availability of biomimetic materials. As (bio)macromolecules are the base of a vast majority of dedicated functions of living organisms it is quite obvious that polymer science has the chance to play a key role in the advent of regenerative therapies. In turn, the challenge of developing biomimetic structures can be expected to substantially influence the field of polymer science as the bidirectional exchange of concepts and technologies with the dynamically evolving disciplines of the life science creates new transdisciplinary areas. Altogether, unraveling and mimicking structures and functions of biopolymers by synthetic architectures will more and more enable the rational design of bioactive materials going beyond the simple imitation of living matter.

The contributions collected in this special volume were selected to mirror the status of research on biomimetic materials in an exciting transition period: a wealth of different strategies has been pioneered and the field is about to further branch out in lively subdisciplines.

A well-established strategy of tissue engineering concerns the use of biodegradable polymers for the generation of cell scaffolds. Among those, natural polyesters from the group of polyhydroxyalkanoates (PHAs) have emerged as particularly promising materials for various applications. Thomas Freier comprehensively describes the characteristics of promising biopolyesters, together with strategies that can be used to adjust the material properties to the clinical requirements and presents examples of potential applications.

Biopolymers of the extracellular matrix (ECM) mark the other edge of the spectrum of the currently used materials. Artificial matrices based on biopolymers isolated from nature were successfully utilized to prepare a variety of bioactive materials capable of supporting desired cell fate transitions to enhance the integration and performance of engineered tissues. Along that line, the review of Tilo Pompe, Katrin Salchert and myself emphasizes recent research to modulate the functionality of ECM biopolymers through their combination with synthetic polymeric materials.

The contribution of Shyni Varghese and Jennifer H. Elisseeff reviews natural and synthetic hydrogels and their use in musculoskeletal tissue engineering. The most appealing feature of hydrogels as scaffolding materials is their structural similarity to ECM and their easy processability under mild conditions. The primary developments in this field comprise formulation of biomimetic hydrogels incorporating specific biochemical and biophysical cues so as to mimic the natural ECM, design strategies for cell-mediated degradation of scaffolds, techniques for achieving in situ gelation which allow a minimally invasive administration of cell-laden hydrogels into the defect site, scaffold-mediated differentiation of adult and embryonic stem cells, and finally, the integration of tissue-engineered "biological implants" with the native tissue.

A promising approach to the generation of biofunctional nanomaterials that may become particularly useful for in vivo tissue engineering strategies concerns the assembly of amphiphilic peptide systems. Xiaojun Zhao and Shuguang Zhang review the current status in this field referring to a classification scheme of the peptide molecules according to their net charge. Recent work dedicated to the formation of nanofibers, bionanotubes and vesicles based on this principle is discussed.

A key aspect of either variant of polymer scaffold is the control of cell adhesion as it is crucial to cellular and host responses to implanted devices, and biotechnological cell culture supports. Emphasizing this aspect the review by Andres J. Garcia focuses on interfaces controlling cell adhesive interactions, highlighting surfaces that control protein adsorption, biomimetic substrates presenting bioadhesive motifs, and micropatterned surfaces. These approaches represent promising strategies to engineer cell-material biomolecular interactions in order to elicit specific cellular responses and enhance the biological performance of materials in a variety of biomedical and biotechnological applications.

Another, similarly important aspect of bioactive materials for regenerative therapies concerns the use of growth factors to promote regeneration of lost or compromised tissues and organs. Claudia Fischbach and David J. Mooney survey the state of the technology to supply growth factors by polymeric systems in a well-controlled, localized, and sustained manner. Design attributes for polymeric systems for VEGF delivery are discussed, and subsequently illustrated in the context of three specific applications: therapeutic angiogenesis, bone regeneration, and nerve regeneration.

We would be extremely delighted if our collection of reviews is of interest to a wide readership and fosters more activity in polymer science dedicated to the development of biomimetic structures. However, by its very nature, the selection involved arbitrariness and I would like to apologize to all those who possibly find their priorities underestimated.

I am deeply obliged to the authors of the following reviews who undertook considerable efforts to summarize a very dynamic field. Beyond that, I would like to thank all colleagues who – although not represented by a review in this issue – significantly contributed to the recent progress in biomimetic polymers and, thus, provided the basis for this survey.

Dresden, May 2006 *Carsten Werner*

Contents

Contents of *Advances in Biochemical Engineering/ Biotechnology*, **Volume 102**

Scaffold Systems for Tissue Engineering

ISBN: 3-540-31944-1

Contents of *Advances in Biochemical Engineering/Biotechnology,* Volume 103

Basics of Tissue Engineering and Tissue Applications

ISBN: 3-540-36185-5

Adv Polym Sci (2006) 203: 1–61
DOI 10.1007/12_073
© Springer-Verlag Berlin Heidelberg 2006
Published online: 7 April 2006

Biopolyesters in Tissue Engineering Applications

Thomas Freier

Smart Biotech Inc., 243 Eglinton Ave. West, Suite 300, Toronto, ON M4R 1B1, Canada
thomas@smart-biotech.com

Abstract Tissue engineering is a rapidly growing interdisciplinary field of research focused on the development of vital autologous tissue through the use of a combination of biomaterials, cells, and bioactive molecules, for the purposes of repairing damaged or diseased tissue and organs. Due to their biocompatibility and biodegradability, as well as their broad range of mechanical properties, natural polyesters from the group of polyhydroxyalkanoates (PHAs) have emerged as promising materials for various tissue engineering applications, including cardiovascular system, nerve, bone, and cartilage repair applications. Thus far, the majority of research on medical applications of

PHAs refers to poly(3-hydroxybutyrate) and its copolymer poly(3-hydroxybutyrate-*co*-3-hydroxyvalerate). In recent years, other PHAs, such as poly(4-hydroxybutyrate) and poly(3-hydroxyoctanoate-*co*-3-hydroxyhexanoate), have drawn increasing attention as viable materials for biomedical applications. Copolymers of poly(3-hydroxybutyrate) and medium-chain-length hydroxyalkanoates offer the advantage of having elastomeric properties. This is of particular importance for engineering of elastomeric tissue, such as in the cardiovascular system, and for providing mechanical stimuli, such as in cartilage repair. This review will describe the characteristics of promising biopolyesters for tissue engineering, together with strategies that can be used to adjust the material properties to the clinical requirements; examples of potential applications will also be presented.

Keywords Biocompatibility · Biodegradability · Mechanical properties · Poly(hydroxyalkanoate)s · Tissue engineering applications

Abbreviations

3HB	3-Hydroxybutyrate
3HH	3-Hydroxyhexanoate
3HV	3-Hydroxyvalerate
4HB	4-Hydroxybutyrate
at-P3HB	Atactic P3HB
BTHC	Butyryltrihexyl citrate
dg-P3HB	Degraded P3HB
ECM	Extracellular matrix
HA	Hydroxyapatite
i.m.	Intramuscular, intramuscularly
i.p.	Intraperitoneal, intraperitoneally
P3HB	Poly(3-hydroxybutyrate)
P3HB-3HV	Poly(3-hydroxybutyrate-*co*-3-hydroxyvalerate)
P3HB-4HB	Poly(3-hydroxybutyrate-*co*-4-hydroxybutyrate)
P3HB-3HH	Poly(3-hydroxybutyrate-*co*-3-hydroxyhexanoate)
P3HO-3HH	Poly(3-hydroxyoctanoate-*co*-3-hydroxyhexanoate)
P3HU	Poly(3-hydroxyundecenoate)
P4HB	Poly(4-hydroxybutyrate)
PCL	Poly(ε-caprolactone)
PDLLA	Poly(D,L-lactide)
PEG	Poly(ethylene glycol)
PGA	Poly(glycolide)
PHA	Poly(hydroxyalkanoate)
PLGA	Poly(D,L-lactide-*co*-glycolide)
PLLA	Poly(L-lactide)
s.c.	Subcutaneous, subcutaneously
st-P3HB	Syndiotactic P3HB
TCP	Tricalcium phosphate
TEC	Triethyl citrate

1
Introduction

Tissue engineering is a rapidly growing interdisciplinary field of research focused on the development of vital autologous tissue, through the use of a combination of biomaterials, cells, and bioactive molecules, for the purposes of repairing damaged or diseased tissue and organs. One of its fundamental concepts is the generation of new functional tissue based on a biodegradable scaffold in the shape of the organ to be replaced. In vitro tissue engineering strategies usually involve seeding the scaffold with autologous cells before implantation. As the cells invade the scaffold and produce extracellular matrix (ECM), thus increasingly lending structure and stability to the tissue, the scaffold is gradually absorbed in vivo. Once absorption is complete, only the newly created functioning tissue remains [1].

Another strategy is in vivo tissue engineering, which can be accomplished by implanting an unseeded scaffold into the damaged region to allow the invasion of new blood vessels, innervation, and deposition of ECM, thereby creating a cell-friendly environment prior to injection of autologous cells into the scaffold [2]. The ultimate in vivo strategy might be guided tissue regeneration, stimulated by biomolecule-loaded unseeded scaffolds that support and control the invasion of (stem) cells from the blood stream and surrounding tissues.

Matrix polymers for tissue engineering must possess certain fundamental properties. A biocompatible material is required that has appropriate mechanical properties, has a suitable surface for supporting cell adhesion and proliferation, that can guide and organize cell growth in the required direction, that can enable tissue to invade and nutrients to be exchanged, and that degrades to nontoxic by-products within desirable periods of time [3].

Biodegradable polymers derived from synthetic aliphatic polyesters, such as poly(glycolide) (PGA), poly(D,L-lactide) (PDLLA), poly(L-lactide) (PLLA), and their copolymers, are widely used as biomaterials in surgical practice [4–7]. At present, bioabsorbable sutures are the predominant medical product made from degradable polyesters. Additionally, suture anchors, orthopedic fixation devices, and various other products made from degradable polyesters are in clinical use. An increasing demand for degradable polyesters in medical use can be expected in the future, with applications in tissue engineering being especially prominent [8–11].

In recent years, biopolyesters from the group of polyhydroxyalkanoates (PHAs) have emerged as promising materials for a variety of medical applications. Potential uses of PHAs can be expected in wound management applications (sutures, skin substitutes, nerve cuffs, surgical meshes, staples, swabs), vascular system applications (heart valves, cardiovascular fabrics, pericardial patches, vascular grafts), orthopedics (scaffolds for cartilage engineering, spinal cages, bone graft substitutes, meniscus regeneration, internal fixation devices), and drug delivery applications [12–17].

The application of polymers as absorbable implants in tissue engineering requires appropriate mechanical, biocompatibility, and degradation properties of the material. However, general specifications of the optimal properties, or a general comparison between different polymers or classes of polymers, cannot be given. Therefore, the question of whether a synthetic polyester or biopolyester is the "better" biomaterial cannot be answered without looking at the specific requirements for a particular application; i.e., the required mechanical properties, host response, and absorption times will eventually guide the selection of the most suitable biomaterial. Similarly, the adjustment of material properties must be individual according to the specific application.

The characteristics of promising biopolyesters for application in tissue engineering, along with strategies to adjust the material properties to the clinical requirements, and examples of potential applications will be outlined in this review article.

2
Biopolyesters and Their Potential in Regenerative Medicine

Polyhydroxyalkanoates (PHAs) are naturally derived polyesters that accumulate as a carbon storage material in a wide variety of bacteria, usually under conditions of limiting nutrients (such as ammonium, sulfate, and phosphate) in the presence of an excess carbon source [14, 18–23]. An imbalanced nutrient supply leads to intracellular storage of excess nutrients. By polymerizing soluble intermediates into insoluble molecules, cells do not undergo alterations of their osmotic state and leakage of nutrients is prevented. Accumulated PHAs form discrete granules that can account for up to 90% of the cell's dry weight [21]. Up to date, approximately 150 hydroxyalkanoate units with different R-pendant groups have been isolated from bacteria [23, 24] (Fig. 1).

Poly(3-hydroxybutyrate), P3HB, is the simplest and most common member of the group of PHAs. Discovered by Lemoigne in the 1920s, its commercial evaluation did not start until the late 1950s. The potential of P3HB for biomedical applications was first suggested in a 1962 patent, which presented the ideas of biodegradable surgical sutures and of films to support tissue healing of injured arteries and blood vessels [25]. In a following patent, prosthetic devices such as support tubes for healing of a severed blood vessel or ureter, as well as support devices for hernia repair, have been described [26].

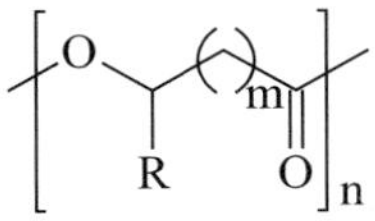

Fig. 1 Chemical structure of PHA homopolymers

It was not until the 1980s that P3HB again became of interest for biomedical research when P3HB tablets for sustained drug delivery were studied [27–29]. Since that time, an increasing number of investigations on the clinical potential of P3HB have been reported, including as implants for bone repair [30], anastomoses tubes and separating films [31], as well as cardiovascular patches [32]. These pericardial patches have subsequently been tested in humans, making this the first reported clinical study of P3HB implants [33]. P3HB conduits and scaffolds have been introduced for the repair of peripheral nerves [34–36] and spinal cord [37], and P3HB patches have been developed for covering damaged tissue of the gastrointestinal tract [38] or injured dura mater [39]. Summarizing these studies, P3HB can be considered to be a polymer with high potential as a degradable implant material [13, 16, 17].

Thus far, the majority of research on medical applications of PHAs refers to P3HB and its copolymer poly(3-hydroxybutyrate-*co*-3-hydroxyvalerate), P3HB-3HV. However, due to their broad range of mechanical and biodegradation properties, other PHAs have been drawing increasing attention for biomedical applications, particularly in cardiovascular tissue engineering. Initially, elastomeric poly(3-hydroxyoctanoate-*co*-3-hydroxyhexanoate), P3HO-3HH, was used to develop scaffolds for repair of blood vessels [40] and heart valves [41]. Subsequently, poly(4-hydroxybutyrate), P4HB, was introduced as a faster degrading alternative to P3HO-3HH and was tested as vascular patches [42], heart valves [43, 44], and vascular grafts [45, 46]. Based on results from these studies, P4HB is regarded as a particularly promising polymer for clinical applications [47]. Furthermore, copolymers such as poly(3-hydroxybutyrate-*co*-4-hydroxybutyrate), P3HB-4HB, and poly(3-hydroxybutyrate-*co*-3-hydroxyhexanoate), P3HB-3HH, have been introduced as scaffolds for tissue engineering, including the repair of cartilage and bone [48, 49]. An example of an unsaturated PHA is PEG-grafted poly(3-hydroxyundecenoate) (P3HU, containing 3-hydroxynonenoate and other unsaturated and saturated side-chains of medium length), which has been suggested as a suitable material for applications in blood contact [50]. The chemical structures of PHAs reviewed in this article are shown in Fig. 2.

The range of properties provided by PHAs is exemplified in Fig. 3 [38, 51–56]. The high crystallinity of the isotactic P3HB leads to stiffness and brittleness, as well as slow hydrolysis in vitro and in vivo. P4HB films are characterized by low stiffness and high elongation at break. The low degree of crystallinity accelerates the in vitro degradation. Interestingly, in vivo degradation of P4HB involves a surface erosion mechanism. P3HO-3HH is an elastomer with a low crystallinity, leading to faster in vitro degradation than P3HB despite the increased hydrophobicity resulting from the longer alkyl side-chains. However, the degradation of P3HO-3HH appears to be slowed down under in vivo conditions.

P3HB copolymers containing more than 20% of 4-hydroxybutyrate [47] or medium chain-length (C6-18) 3-hydroxyalkanoate units [57], as well as

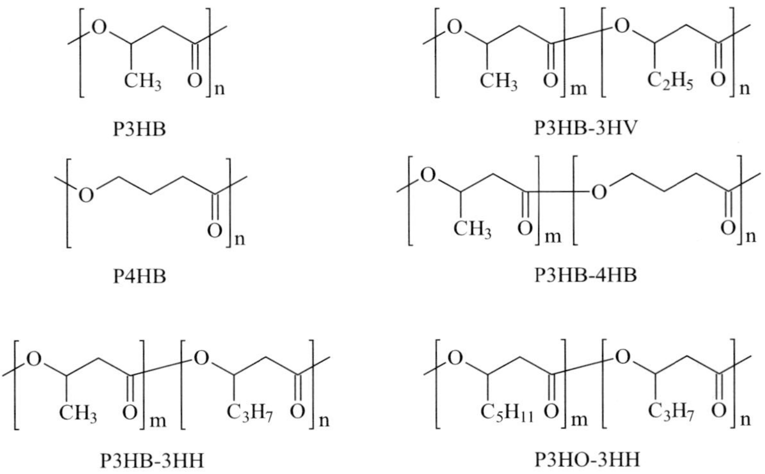

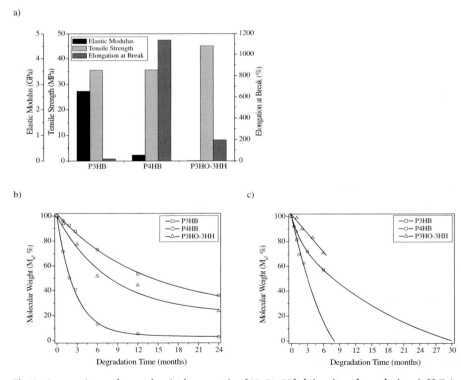

Fig. 2 PHAs tested in tissue engineering applications

Fig. 3 Comparison of **a** mechanical properties [38, 51, 52], **b** in vitro degradation (pH 7.4, 37 °C) [38, 51, 53, 54], and **c** in vivo degradation (s.c., mice or rats) [53, 55, 56] of P3HB, P4HB, and P3HO-3HH films

medium chain-length PHA homopolymers (e.g., P3HO-3HH [58] or cross-linked P3HU [50]) offer the advantage of having elastomeric properties. This is of particular interest in tissue engineering applications since many tissues in the body have elastomeric properties. For example, cardiovascular tissue engineering requires a scaffold that can sustain and recover from cyclic deformations without irritation of the surrounding tissue. The development of an elastomeric scaffold is therefore highly desirable. Moreover, mechanical stimuli promote the formation of functional tissue, for example in cardiovascular or cartilage tissue engineering, and allow for gradual stress transfer from the degrading synthetic matrix to the newly formed tissue. Most of the biodegradable polyesters currently used in tissue engineering (such as PGA, PDLLA, and their copolymers) undergo plastic deformation and fail when exposed to long-term cyclic strain, thereby limiting their efficacy in engineering elastomeric tissue [59].

3
Poly(3-hydroxybutyrate)
and Poly(3-hydroxybutyrate-*co*-3-hydroxyvalerate)

P3HB is the classic, most extensively studied and characterized PHA. It is produced by a large number of microorganisms, including Gram-negative and Gram-positive aerobic and photosynthetic species, lithotrophs or organotrophs [18]. A glucose-utilizing mutant of *Alcaligenes eutrophus* can accumulate up to 80% of P3HB with glucose as the carbon source. By addition of propionic acid to the medium, P3HB-3HV is produced as a random copolymer, with the comonomer ratio dependant on the ratio of propionic acid to glucose. This technology has been scaled-up, and industrial processes for the synthesis of large quantities of P3HB-3HV (Biopol) have been developed [60]. However, one of the major drawbacks for the broad utilization of P3HB or P3HB-3HV in medicine is the limited supply of these polymers of medical grade. Furthermore, there is currently no Drug Master File submitted to the FDA, and a medical device made of P3HB or P3HB-3HV has not yet been clinically approved.

3.1
Mechanical Properties

P3HB as a natural thermoplastic polyester has mechanical properties comparable with those of synthetically produced degradable polyesters such as the polylactides [61]. The relatively high brittleness of the crystalline natural isotactic P3HB is of disadvantage in tissue engineering applications but can be overcome by copolymerization and incorporation of PHA components such as 3-hydroxyvalerate (3HV) [60], 4-hydroxybutyrate (4HB) [62], or

3-hydroxyhexanoate (3HH) [63]. P3HB-4HB and P3HB-3HH copolymers will be described in Sects. 4 and 5 of this article. The effect of increasing 3HV content on the properties of P3HB-3HV is shown in Table 1 [64]. Crystallinities of P3HB-3HV copolymers are high over the whole composition range due to isodimorphism, i.e., cocrystallization of the two monomer units in either of the homopolymer crystal lattices [65]. Molecular weight is one of the major factors governing the mechanical properties of a polymer. Thus, the tensile strength of P3HB-3HV has been reported to rapidly decrease below molecular weights of about 100 000 [66] to 150 000 [67].

The mechanical properties of P3HB can also be improved by addition of plasticizers [68–72] (Fig. 4a). For example, citric acid esters, which are considered to be nontoxic [73, 74], were demonstrated to be effective plasticizers for P3HB-3HV [75–78]. However, triethyl citrate, which is commonly used for plasticization, is highly water soluble resulting in fast leaching out of the polymer matrix under physiological conditions leading to embrittlement of initially plasticized polymer films [38, 79]. Recently, it has been shown that esters of anti-inflammatory drugs such as salicylic acid, acetylsalicylic acid, and ketoprofen have plasticizing effects on P3HB films comparable to those of citric acid esters [71].

Blending with other degradable polyesters, such as atactic P3HB (at-PHB) [38, 80–83], syndiotactic P3HB (st-PHB) [81, 82, 84], poly(ε-caprolactone) (PCL) [85–87], poly(6-hydroxyhexanoate) [88], P3HB-3HH [89, 90], and P3HO-3HH [52], has been shown to increase the flexibility and elongation at break of P3HB or P3HB-3HV. Some examples of P3HB blends are given in Fig. 4b. In contrast, addition of polyesters commonly used in medical applications, such as PDLLA, PLLA, poly(D,L-lactide-co-glycolide) (PLGA), or poly(p-dioxanone), does not improve the mechanical properties of P3HB due to lack of miscibility.

Polymer films with improved mechanical properties can also be obtained by hot-drawing of ultrahigh molecular weight P3HB [91] or P3HB/PLLA blends [92]. If unmodified, films made from ultrahigh molecular weight P3HB

Table 1 Thermal and mechanical properties of P3HB and P3HB-3HV [64]

Polymer	T_g [°C]	T_m [°C]	Elastic modulus [GPa]	Tensile strength [MPa]	Elongation at break [%]
P3HB	9	175	3.8	45	4
P3HB-11%3HV	2	157	3.7	38	5
P3HB-20%3HV	− 5	114	1.9	26	27
P3HB-28%3HV	− 8	102	1.5	21	700
P3HB-34%3HV	− 9	97	1.2	18	970

a)

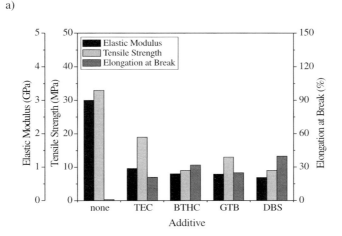

b)

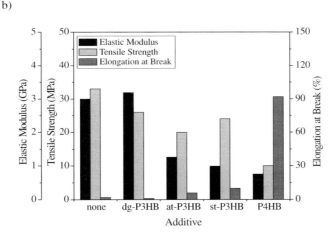

Fig. 4 a Mechanical properties of solution-cast P3HB films plasticized with 30% each of triethyl citrate (TEC), butyryltrihexyl citrate (BTHC), glycerin tributyrate (GTB), and dibutyl sebacate (DBS) [71, 72]. **b** Mechanical properties of solution-cast P3HB films blended with 30% each of degraded P3HB (dg-P3HB, M_w = 3000), at-P3HB, st-P3HB, and P4HB [38, 72]

(M_w 11 million) have comparable mechanical properties with those of common high molecular weight P3HB (M_w 500 000 to 1 million). Low molecular weight P3HB shows increasing brittleness with decreasing molecular weight, and mechanically stable films cannot be prepared below a molecular weight of 50 000. Films can still be prepared from mixtures of high and low molecular weight P3HB showing mechanical properties comparable with those of unmodified high molecular weight P3HB (Fig. 4b) [38].

P3HB-based composites, which are of interest in bone tissue engineering, were found to have an increasing stiffness with an increasing content of hydroxyapatite (HA) [93, 94] or tricalcium phosphate (TCP) [95].

3.2
Biocompatibility

Material–tissue interactions are best described by the term biocompatibility. Generally defined, biocompatibility is "the ability of a material to perform with an appropriate host response in a specific application" [96]. This implies that any material placed into a body will not be inert but will interact with the tissue. The biological response of a material is basically dependent on three factors: the material properties, the host characteristics, and the functional demands on the material. Therefore, the biocompatibility of a material can only be assessed on the basis of its specific host function and has to be uniquely defined for each application.

Toxicity Testing

P3HB has been found to be an ubiquitous component of the cellular membranes of animals [97]. The resulting presence of relatively large amounts of low molecular weight P3HB in the human blood, as well as the fact that the degradation product, 3-hydroxybutyric acid (3HB), is a common metabolite of all higher living beings, serve as evidence for the nontoxicity of P3HB [20].

Toxicity testing according to USP XXII and ISO 10993 revealed that P3HB is suited for use as an implant material. The subcutaneous, intraperitoneal, and intravenous eluate testing did not result in any significant reactions in rabbits, mice, or guinea pigs, and no febrile reactions were observed during the pyrogen test. Histocompatibility was demonstrated in the implantation study in rabbits (Schmitz KP, personal communication).

P3HB did not cause any inflammation in the chorioallantoic membrane of the developing egg [98]. P3HB-5%3HV was nontoxic in the bacterial bioluminescence test over a period of 16 weeks [99]. P3HB-3HV (7%, 14%, 22% 3HV) films were found to elicit only a mild toxic response in the direct contact or agar diffusion tests [100]. However, P3HB-22%3HV extracts in saline provoked a noticeable hemolytic reaction [101].

Cell Culture Studies

Mouse fibroblast cell-lines are relatively unaffected by small changes in cell culture conditions and are therefore commonly used to assess and compare the cell compatibility of biomaterials in vitro. For example, it was reported that NIH 3T3 mouse fibroblasts remained highly viable on P3HB and P3HB-3HV (15%, 28% 3HV) films [102]. In another study, L929 mouse fibroblasts showed a better viability on P3HB surfaces than on PLLA [94]. A good cell compatibility of P3HB films has also been concluded from experiments using

Chinese hamster lung (CHL) fibroblasts [103]. Typical results from cell compatibility tests of P3HB films using L929 mouse fibroblasts are shown in Figs. 5 and 6 [72]. Melt-spun P3HB fibers [104, 105] also support L929 mouse fibroblast adhesion, as shown in Fig. 6c.

Canine anterior cruciate ligament fibroblasts cultured in highly porous P3HB-9%3HV scaffolds sustained a cell growth similar to that observed in collagen sponges [106]. On the other hand, human scoliotic fibroblasts isolated from spinal ligaments exhibited low proliferation rates on P3HB-3HV (7%, 14%, 22% 3HV) surfaces, independent of the copolymer composition [107]. P3HB-8%3HV was found to be slightly more compatible than P3HB-12%3HV in terms of L929 fibroblast proliferation, cytotoxic effect, and cytokine production [108].

A limited cell compatibility of plasticized P3HB-3HV was observed in a study using NIH 3T3 mouse fibroblasts, and was attributed to leaching of plasticizers [109]. Thus, the potential toxicity of leachables such as plasticizers or other additives has to be considered when conducting cell culture experiments. Additionally, removal of remaining solvent is crucial if polymer films are prepared by solution-casting, which is the most widely used method for sample fabrication in cell culture studies. It has been shown that storage of solution-cast polymer films even under vacuum does not allow for complete chloroform removal (Fig. 7), so that solvent may be released from the polymer film during cell culture experiments.

The relatively low surface wettability of P3HB solution-cast film surfaces (contact angle 68° [94, 110]) has been discussed as a limiting factor for cell attachment and growth. Strategies to improve cell compatibility of P3HB include methods to increase the surface wettability (decreasing water con-

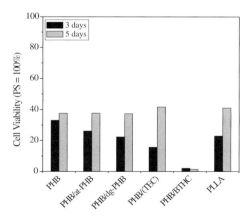

Fig. 5 Cell viability of L929 mouse fibroblasts after 3 and 5 days of culture on solution-cast P3HB films, compared to that of P3HB films modified with 30% at-P3HB, dg-P3HB, TEC (leached), and BTHC; PLLA films for comparison [72]

a)   b)

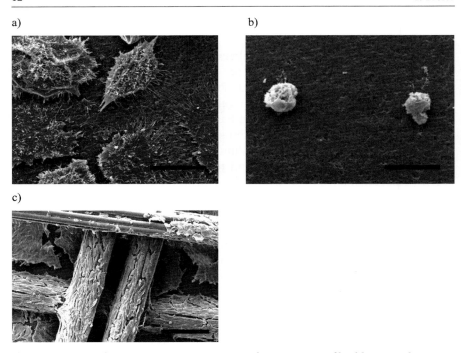

c)

Fig. 6 Scanning electron microscopy images of L929 mouse fibroblast attachment to **a** P3HB films, and **b** P3HB films plasticized with 30% BTHC after 5 days of culture [72]. P3HB/BTHC films show limited cell adhesion, and cells exhibit a spherical, nonviable morphology. The low cell viability of these films was confirmed by the quantitative analysis (see Fig. 6). *Scale bar* 20 μm. **c** L929 mouse fibroblast attachment to P3HB fibers after 9 days of culture [72]. *Scale bar* 200 μm

tact angle) such as carboxyl ion implantation [110] or oxygen plasma treatment [111]. Introduction of oxygen-based functionalities on the surface of P3HB-15%3HV films, either by corona-discharge treatment or treatment with a mixture of perchloric acid and potassium chlorate, resulted in decreasing water contact angles and increasing mouse NIH 3T3 fibroblasts adhesion and proliferation [112]. A limited adherence of L929 mouse fibroblasts on unmodified P3HB films has been reported in another study. However, a strong improvement of cell adherence and growth could be observed on UV-irradiated films and after fibronectin coating [113], as well as after surface modification by ammonia plasma treatment [114].

Low viability of L929 fibroblasts on unmodified P3HB films, but strongly improved viabilities after surface hydrolysis using lipases or alkaline solution, have been reported [89, 115, 116]. In experiments with P3HB porous scaffolds, an increasing growth of L929 cells could be observed after surface treatment with lipase but a decreasing cell growth was found after coating with hyaluronic acid despite significantly decreased water contact angles. It

a)

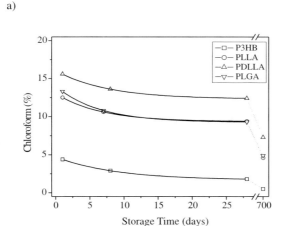

b)

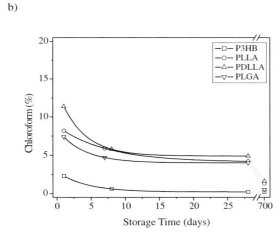

Fig. 7 Chloroform content of solution-cast polyester films after storage **a** at normal pressure (23 °C) and **b** under vacuum (40 °C) [79]

was concluded that an appropriate balance of hydrophilic and hydrophobic surface properties is required for appropriate protein adsorption and cell attachment [117]. Surface modification of P3HB-5%3HV films with hyaluronic acid or chitosan after ozone treatment and acrylic acid grafting has also been examined. Enhanced cell attachment and proliferation of L929 fibroblasts have been observed after immobilization of hyaluronic acid. Chitosan grafting leads to more cell attachment to the film surfaces but less proliferation in comparison to hyaluronic acid. It was shown that fibroblast attachment increases but proliferation decreases with increasing surface density of amine groups [118].

In order to assess the biocompatibility of degradable polymers, it is necessary to study not only the polymer properties but also those of the low molecular weight degradation products. 3HB, the ultimate degradation product of P3HB, has been found to be biocompatible in tests with CHO-K1 fibroblasts [119]. Other studies included short-chain P3HB oligomers. For example, the incubation of P3HB oligomers with hamster V79 fibroblasts and marine melanoma B16(F10) cells did not affect the cell viability [120]. Furthermore, particles of short-chain P3HB were found to cause dose-dependent cell damage in macrophages but not in fibroblasts. Therefore, controlling the degradation rate and thus concentration of degradation products was concluded to be important for controlling the biocompatibility. Cocultures of liver cells (Kupffer cells and hepatocytes) were not affected by treatment with this material [121].

A murine monocytes-macrophages cell line was used as a model to study foreign-body response and phagocytosis capability. The adhesion and proliferation of these cells increased after alkaline surface hydrolysis of P3HB-8%3HV films. Pretreatment of the unmodified polymer with collagen had a repulsive effect on the cells, which disappeared on the hydrolyzed polymer, while fibronectin promoted cell adhesion on both surfaces [122].

Despite these screening studies to assess the cell compatibility of P3HB-based materials in general (including suitable methods for processing, modification, purification, and sterilization), a number of studies have been conducted to study the potential and biocompatibility of P3HB in tissue engineering applications using different tissue-specific cell types, such as osteoblasts, chondrocytes, and vascular cells.

For example, the biocompatibility of P3HB has been assessed by studying the structural organization of cellular molecules involved in adhesion using osteoblastic and epithelial cell lines. Both cell lines exhibited a rounded cell shape due to reduced spreading on the polymer surface. The interactions between matrix proteins and the actin cytoskeleton mediated by integrins were found to be impaired, including the colocalization of fibronectin fibrils with actin filaments. Moreover, the cell morphology was modified showing larger lateral extensions in the cell–cell contacts [123]. Osteoblast compatibility has also been tested after seeding of rabbit bone marrow cells on P3HB films and more cells could be found on P3HB than on PLLA surfaces [94]. P3HB porous scaffolds showed slightly higher osteoblast viability, but comparable alkaline phosphatase production to that of PLLA samples [49]. P3HB-3HH copolymer matrices showed superior cell compatibility in both studies (see Sect. 5.2). Addition of hydroxyapatite to the P3HB scaffolds resulted in enhanced viability and alkaline phosphatase activity [124]. Cell culture studies with primary human osteoblasts showed very limited cell attachment and proliferation on P3HB-7%3HV in comparison to P3HB surfaces [125]. Surface modification of P3HB-8%3HV porous scaffolds by oxygen plasma treatment resulted in more viable rat bone marrow osteoblasts and increased alkaline phosphatase activity in comparison to unmodified materials [126, 127]. Covalent immobilization of collagen I onto the surface of P3HB-8%3HV films after ozone treatment and methacrylic acid grafting also enhanced bone cell growth, as tested with mouse osteoblastic and rat osteosarcoma cells [128].

Chondrocytes derived from rabbit articular cartilage were seeded on P3HB films [129] and porous scaffolds [48, 89, 130, 131]. Cells attached to P3HB films secreted both collagen II and collagen X, which are major cartilage-specific ECM proteins, indicating maturational differentiation of chondrocytes into cartilage [129]. However, collagen II expression was low on P3HB in comparison to PLLA scaffolds [131]. Enhanced cell adhesion and growth, as well as more collagen II synthesis, could be obtained after addition of P3HB-3HH to the P3HB matrices (see Sect. 5.2) [48, 89, 90, 129–131]. In another study, P3HB-9%3HV porous matrices incubated with ovine chondrocytes showed lower cell densities in comparison with collagen sponges [132]. However, electrospun P3HB-5%3HV nanofibrous mats promoted chondrocyte attachment when compared with polymer films [133].

Other cell types tested included mammalian or human epithelial cells, which showed little or no cell adhesion on P3HB fiber-based "wool". However, surface treatment with acidic or alkaline solutions promoted cell proliferation on these fibers [134, 135].

Cells of the human respiratory mucosa (fibroblasts and epithelial cells) have been incubated with P3HB films as a potential replacement matrix of respiratory mucosa after surgical resections. However, while PLLA and collagen supported cell growth, P3HB showed cell growth only after surface modification by intense ammonia plasma treatment. No differentiation of epithelial cells with beating cilia could be found on all materials during the 6-week study [136].

The attachment rate of human retinal pigment epithelium cells was higher on P3HB-8%3HV films modified by oxygen plasma than on unmodified films due to increasing hydrophilicity and decreasing surface roughness. The cells were grown to confluency as an organized monolayer suggesting P3HB-8%3HV as a potential temporary substrate for subretinal transplantation to replace diseased or damaged retinal pigment epithelium [137].

An excellent biocompatibility of P3HB was found in cell culture studies with Langerhans cells [138]. The biocompatibility of plasma-modified P3HB [139] and P3HB-9%3HV [140] was assessed by analysis of insulin secretion of Langerhans cells cultured on the polymer films. The following order of insulin secretion was found during the first 2 days of cell incubation: oxygen > allylamine > allylalcohol > nontreated > argon > water plasma treatment [140].

Cell culture experiments using mouse liver cells (endothelial cells and hepatocytes) grown on P3HB and P3HB-3HV films (15%, 28% 3HV) suggested lack of cytotoxicity of the highly purified materials tested [102]. The adhesion of human endothelial cells on P3HB films could be controlled by plasma treatment introducing positive (NH_3 plasma) or negative (H_2O plasma) surface charge, which was explained by modulated anchorage and conformational changes of adsorbed fibronectin [141].

In Vivo Studies

A number of in vivo studies have shown very mild tissue reactions after implantation of P3HB, which were comparable to those of other polymers in medical use. For example, an excellent in vivo biocompatibility of P3HB has been reported after s.c. implantation in mice [29]. Nonwoven P3HB patches tested as pericardial substitutes in sheep appeared to be slowly phagocytosed by polynucleated macrophages without any other kind of inflammatory cells, except for a small number of lymphocytes. The number of macrophages surrounding the polymer particles decreased with the absorption of the polymer [142]. Histologically, a dense collagen layer similar to that in native pericardium was found on the epicardial side of the patch. A thin fibrotic layer surrounding the patch disappeared along with the macrophages when the patch was absorbed but the regenerated pericardial tissue remained [143]. A mild inflammatory response to the nonwoven P3HB patch material similar to that reported for the pericardial substitutes appeared after closure of an atrial septal defect [144] and enlargement of the right ventricular outflow tract in sheep [145].

P3HB and P3HB-3HV (5.5%, 9%, 19%, 22% 3HV) films manifested good tissue tolerance when implanted s.c. in mice for up to 6 months. No acute inflammation, abscess formation, or tissue necrosis was observed. At 1 month after implantation, implants were surrounded by a fibrous, vascularized capsule consisting primarily of connective tissue cells. A mild inflammatory reaction was manifested by the presence of mononuclear macrophages, foreign-body cells, and lymphocytes. The number of inflammatory cells increased with increasing valerate content in the copolymer. Thus, inflammation was most pronounced for P3HB-22%3HV. Three months after implantation, the fibrous capsule had thickened due to an increase in the amount of connective tissue and a few collagen fiber deposits. A substantial decrease in inflammatory cells was observed at this time, but inflammation still remained more pronounced for P3HB-3HV with a higher content of valerate units. After 6 months of implantation, the number of inflammatory cells had further decreased and the fibrous capsule, consisting mainly of collagen fibers, had thinned. Within 3 months of implantation a slightly stronger tissue reaction to P3HB than to PDLLA or PLLA was observed in this study and attributed to low molecular weight components and impurities leaching out of the polymer samples. At 6 months, the tissue response to the implants was similar for all these types of polymers [55].

An increasing inflammatory response with increasing 3HV content was also observed for P3HB-3HV (8%, 12% HV) films s.c. implanted in rats for up to 12 weeks [108, 146]. Few differences between P3HB-3HV copolymers in terms of tissue response have been found after i.m. implantation of P3HB-3HV (7%, 14%, 22% HV) in sheep for up to 90 weeks. Acute inflammatory reactions significantly decreased with time and no abscess formation or tissue necrosis were reported in this study [100]. The tissue response to P3HB and P3HB-3HV (15% HV) fibers implanted i.m. in rats was characterized by a short acute inflammation period (up to 2 weeks) followed by the formation of a fibrous capsule of less than 200 µm thickness during weeks 4 to 8, which was reduced to 40–60 µm after 4–6 months. Forty eight weeks after surgery, the fibrous capsule surrounding the implants was minimal. Mono- and polynuclear macrophages were still abundant at this time. The tissue response to P3HB and P3HB-3HV fibers was similar in terms of inflammatory reaction and fibrous capsule formation. There were no adverse reactions, such as suppurative inflammation, necrosis, calcification, and malignant tumor formation for up to 48 weeks after implantation [147, 148].

P3HB discs implanted for 3 months in the peritoneum of rats showed the presence of a thin and poorly adherent fibrous capsule that contained no inflammatory cells. P3HB was assessed to be biocompatible because the formed capsule was porous and ensured communication between the polymer and the biological fluids. Perfluorohexane plasma-modified P3HB discs were surrounded by a nonporous capsule indicating a slight decrease in the surface biocompatibility [139].

P3HB samples implanted s.c. and i.p. in rats were found to be tissue compatible without inflammation reaction or tumor formation. After 1 year of implantation the fibrous capsule surrounding the s.c. implant was about 30% thicker than that surrounding the i.p. implant. A small number of phagocytes indicated the polymer resorption process at that time [149].

No inflammatory reactions resulted from the i.m. injection of P3HB or 3HB in rats, indicating a good in vivo biocompatibility [119]. P3HB patches implanted onto the rat stomach showed no significant inflammatory response as confirmed by analysis of cytokine production. A group of mRNAs encoding pancreatic enzymes was transiently present in tissue surrounding the patch material [150] 1–2 weeks after implantation. The amount of mRNA of the inflammation marker C-reactive protein (CRP) was also found to transiently increase [151].

P3HB in bone contact causes a strong initial cellular reaction with slight or no inflammation [152]. P3HB bone screws showed an optimal tissue compatibility [153]. Osteosynthesis plates and screws made of P3HB were highly compatible without induction of immunologic or inflammatory reactions [154]. A very fast healing and formation of new bone substance could be achieved with P3HB/HA composites [93]. The interface between the composite and bone was physically and biochemically active over a 6-month period of implantation into the condyles of rabbits. Bone bonding to these composites occurred by degradation of the P3HB matrix, which led to the formation of new crystallites between the parent HA particles in the P3HB/HA composite, as well as at the surface of the parent HA particles [155].

With respect to the mechanical properties and the tissue response, P3HB-20%HV films were more suitable than PLLA, PDLLA, and PCL to separate mucoperiostum and bone in a dog model for closure of palatal defects [156]. In contrast, P3HB-3HV reinforced with polyglactin was not useful as an occlusive barrier over dental implants in dogs, since the material prevented bone healing due to an increased inflammatory reaction [157].

P3HB-22%3HV coated onto a tantalum stent implanted for 4 weeks in the porcine coronary artery induced a marked inflammatory and foreign body response, thrombosis as well as extensive fibromuscular proliferation leading to eccentric stenosis [158]. Intense inflammatory reactions and proliferations, thrombosis, and in-stent lumen narrowing have also been reported after implantation of P3HB stents (plasticized with 30% TEC) into the rabbit iliac artery for up to 30 weeks. The polymer degradation process was suggested to be the main reason for the significant chronic inflammation induced by these stents [159]. Another explanation might be the fast leaching of the water-soluble plasticizer [79], together with polymer crystallization induced by the laser-cutting in the stent manufacturing process [160] leading to polymer stiffness and brittleness, which may cause the tissue irritation and early stent rupture observed in the study.

Blood Compatibility

In vitro studies have confirmed that P3HB is a polymer with a high blood compatibility [139]. For example, a good thromboresistance of the P3HB surface has been concluded from the adsorption and desorption characteristics of albumin and fibrinogen, the latter playing a major role in stimulating thrombus formation [161]. Furthermore, the blood compatibility of P3HB, P3HB-9%3HV, and P3HB-22%3HV films has been compared by studying the adsorption of these proteins. An increasing fibrinogen adsorption has been observed with increasing 3HV content (increasing hydrophobicity) while albumin adsorption decreased [162].

The protein adsorption that precedes and probably determines the subsequent coagulation process was also evaluated on unmodified P3HB films and on films modified by perfluorohexane plasma. It was ascertained that the adsorption of fibrinogen strongly decreased on the plasma-treated P3HB films, resulting in prolonged blood coagulation times in vitro. Perfluorohexane plasma modification leads to very smooth and hydrophobic surfaces [138, 139]. Surface roughness was found to be more significant for protein adsorption than surface energy [139]. The same was concluded after alkaline treatment of P3HB-8%3HV films leading to strong surface erosion and increasing area accessible to proteins [122].

The release of molecules of the inflammation cascade was evaluated in vitro after contact of P3HB films with fresh human blood. The activation of the cellular coagulation (activation of thrombocytes with release of platelet factor 4 or β-thromboglobulin) and plasmatic coagulation (release of pro-thrombin fragment F1+2) was observed, the latter, however, only to a small extent [163]. The complement activation of human serum due to P3HB contact was demonstrated by a significant increase of the C3a-desArg con-centration [164]. However, no significant activation of the coagulation and complement systems by P3HB films have been observed in subsequent studies (Fig. 8) [72]. The conflicting results might be attributed to different degrees in the purity of the polymer used in both sets of experiments.

The importance of polymer purity on the blood compatibility of P3HB and P3HB-3HV (4%, 18% 3HV) has recently been investigated. It has been

a)   b)

c)

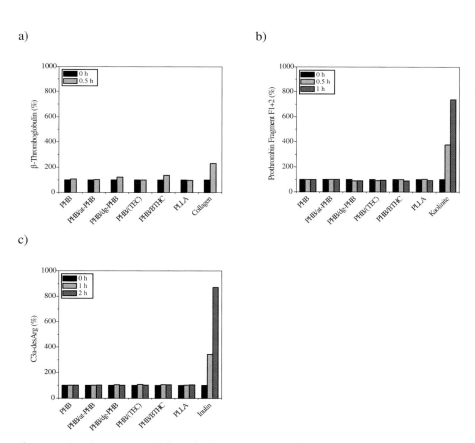

Fig. 8 In vitro hemocompatibility of solution-cast P3HB films, compared to that of P3HB films modified with 30% each of at-P3HB, dg-P3HB, TEC (leached), and BTHC; PLLA films for comparison. Activation of **a** cellular coagulation (release of β-thromboglobulin, control = collagen), **b** plasmatic coagulation (release of prothrombin fragment F1+2, con-trol = kaolinite), and **c** complement system (release of C3a-desArg, control = inulin) [72]

shown that additional purification steps to remove residual bioactive components from the polymeric material (such as lipopolysaccharides (endotoxins) stemming from bacterial cell walls) significantly reduce complement activation. Interestingly, the rate of complement activation by the polymers tested decreased with increasing amounts of 3HV, either before or after additional purification [165].

Other P3HB-based materials tested included P3HB/PEG blends, which exhibited increasing blood coagulation times and less platelet adhesion with increasing amount of PEG [103]. In another study, P3HB-3HV (5% HV) films immobilized with hyaluronic acid had less protein (albumin, fibrinogen) adsorbed and longer coagulation times than unmodified films. Immobilization with chitosan had the opposite effect [118].

In vivo, a complete coverage of the surface of nonwoven P3HB patches with endothelial-like cells without any platelet or fibrin adhesions was reported from a 12-month study of closure of an atrial septal defect in sheep [144]. No platelet aggregates were seen and fibrin rarely observed on the patch surface in a 24-month study on the right ventricular outflow tract of sheep [145].

However, severe thrombosis was observed in vivo when testing P3HB-22%3HV-coated tantalum stents in pigs [158]. The hydrophobic nature of the polymer surface supporting protein adsorption, followed by platelet adhesion and subsequent thrombi formation, was given as a possible reason for the stent failure [166]. This is in accordance with in vitro results mentioned above [162]. Based on these experiments, P3HB-3HV was assessed to be less suited for implants that are in immediate and extended blood contact [166]. Severe thrombosis, including complete thrombotic occlusion, also occurred after implantation of P3HB stents but not with tantalum controls in a rabbit study [159]. It must be considered that the polymer stents used in this study were plasticized with 30% TEC, which can leach out quickly under physiological conditions leaving behind a more porous polymer structure that may increase protein adsorption.

3.3
Biodegradability

In Vitro Degradation

In addition to suitable mechanical and biocompatibility properties, a temporary implant material needs to degrade within clinically reasonable time periods. In vitro degradation studies on P3HB films in buffer solution (pH 7.4, 37 °C) showed no mass loss after 180 days, but a decrease in molecular weight starting after an induction period of about 80 days [167]. This induction period was attributed to the time required for water to penetrate the polymer matrix. The degradation mechanism was examined in an accelerated test at 70 °C. It was concluded that the hydrolysis of microbial polyesters proceeds in two steps. First, there is a random chain scission both in the amorphous and crystalline regions of the polymer matrix associated with a decrease in the molecular weight with unimodal distribution and relatively narrow polydispersity. Simultaneously, an increase in crystallinity occurs that is attributed to the crystallization of chain fragments in the hydrolyzed amorphous regions. When the molecular weight of the degrading polymer reaches a critical low M_n of about 13 000, mass loss starts as the second step [167].

This in vitro degradation profile is typical for a bulk-degrading polymer and can also be found in synthetic polyesters in medical use, such as PGA, PLGA, PDLLA, and PLLA. Thus, significant mass loss of P3HB will be observed only after the prolonged period of time necessary for the molecular weight to reach the critical lower limit. Therefore, the degradation profile can hardly be predicted solely by determination of mass loss but should preferably be based on the molecular weight analysis.

Unmodified P3HB is a relatively slowly degrading polymer. However, an acceleration of the degradation rate is possible by simple modifications, such as the addition of polymers or plasticizers. One strategy is the addition of hydrophilic or amorphous polymers, which enhance the water absorption into the polymer bulk and accelerate its hydrolysis. For example, the water content was found to be higher in P3HB/PDLLA than in P3HB/PCL blends [168]. Amorphous at-P3HB degrades faster than natural isotactic P3HB and can also be used as blend component [169]. Another strategy is the leaching of water-soluble additives, which leads to a higher polymer surface area, as discussed for P3HB and glycerin derivatives [170]. The addition of PDLLA oligomer or PEG as amorphous and leachable additives has also been shown to accelerate the hydrolysis of, for example, P3HB-12%3HV [171].

The in vitro degradation of solution-cast films of P3HB and modifications of P3HB expected to influence the degradation time has been studied (Fig. 9a). The molecular weight of unmodified P3HB decreased to one half of its original value after 1 year of storage in buffer solution (pH 7.4, 37 °C). An accelerated degradation could be achieved by blending with 30% at-P3HB, while addition of low molecular weight (predegraded) dg-P3HB had no influence due to its high crystallinity. Leaching of a water-soluble additive (TEC) led to a slight initial acceleration of the P3HB degradation. In contrast, a deceleration of the degradation rate was observed after addition of a hydrophobic plasticizer (BTHC) [38]. Slightly enhanced hydrolysis rates could be achieved by increasing the amount of at-P3HB in the P3HB/at-P3HB blends [39] (Fig. 9b).

An accelerating effect of pancreatin containing a mixture of enzymes (including lipase, amylase, α-chymotrypsin, trypsin and protease) on the hydrolysis of P3HB, but not PLLA, has been observed in this study. The degradation rate of P3HB was accelerated about threefold in comparison to simple hydrolysis [38]. There are only a few other reports on this topic in the literature. For example, microparticles made of a PHB-10.8%HV/PCL (80/20)

a)

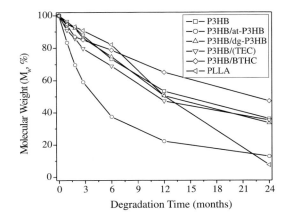

b)

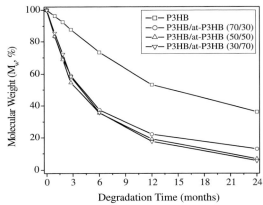

Fig. 9 a In vitro degradation (pH 7.4, 37 °C) of solution-cast P3HB films, compared to that of P3HB films modified with 30% each of at-P3HB, dg-P3HB (M_w = 3000), TEC (leached), and BTHC; PLLA films for comparison. After 2 years in buffer solution, significant mass loss was observed only for films made of P3HB/at-P3HB (12%) and P3HB/BTHC (10%) [38]. **b** Effect of increasing amount of at-P3HB in P3HB/at-P3HB blends on the in vitro degradation (pH 7.4, 37 °C) of solution-cast films. There was no mass loss of P3HB films after 2 years in buffer solution while mass loss of P3HB/at-P3HB films increased with increasing amount of at-P3HB (30/70: 12%, 50/50: 18%, 70/30: 29%) [39]

blend showed accelerated mass loss with bulk and surface erosion after incubation in newborn calf serum or pancreatin solution compared to plain buffer solution. This was attributed to the effects of enzyme activity over and above simple ester hydrolysis [172]. The potential participation of enzymes found in the human body in the hydrolysis of P3HB was also discussed in another study in which a mass loss of more than 10% in 41 days was observed after incubation of polymer films in the presence of lysozyme [103]. Microbial lipases, on the other hand, had no effect on the P3HB degradation [170, 173]. The occurrence of low molecular weight P3HB in the human body may suggest that specific enzymes are involved in the polymer synthesis and its depolymerization in vivo. Moreover, it is worth noting that a Ser-His-Asp triad constitutes the active center of the catalytic domain of both P3HB depolymerase [174] and pancreatic lipase [175]. The serine is part of the pentapeptide Gly-X_1-Ser-X_2-Gly, which is located in all known P3HB depolymerases as well as lipases, esterases, and serine proteases [174]. However, the occurrence of P3HB-specific enzymes in the human body and their contribution to the hydrolysis of implanted P3HB needs still further clarification.

The hydrolytic degradation of P3HB-3HV copolymers with low hydroxyvalerate content (up to 20%) has been extensively studied. A comparison with the P3HB homopolymer is possible because the low 3HV content has only little influence on the hydrolysis rate [176] while the molecular weight, crystallinity, porosity, and any additives are of significant importance. P3HB-3HV samples showed an increased degradation rate with decreased initial molecular weights [176–178]. The degree of crystallinity depends on the processing so that an acceleration of P3HB degradation is in the order of decreasing crystallinity: injection molding < melt pressing < solution casting [177]. P3HB samples produced by cold compression degrade even faster than the solution cast ones [177, 178]. The addition of hydrophilic polysaccharides [179] resulted in a strong acceleration of the hydrolysis rate of P3HB-3HV, whereas the addition of hydrophobic PCL [180] showed only little influence on the hydrolysis rate. The degradation of P3HB and P3HB-3HV films can also be accelerated by incorporation of basic molecules [181, 182].

P3HB-5%3HV samples did not show any mass loss after 16 weeks in buffer solution (pH 7.4, 37 °C), while the decreasing viscosity indicated a molecular weight reduction [99]. The degradation times of P3HB and high molecular weight PLLA, as well as PCL, were comparable in this study. P3HB-7%3HV showed, after an initial period, a continual decrease in molecular weight to nearly half of its initial value after 2 years in buffer solution (pH 7.4, 37 °C) [183]. However, using the same material, a continual decline in the molecular weight without induction period was detected in another study [184]. P3HB-3HV (7%, 14%, 22% 3HV) tested in an accelerated degradation test (pH 7.4, 70 °C) for up to 220 days showed increasing mass loss with increasing 3HV content [100].

Films made of P3HB-6%3HV showed a two-step hydrolysis (pH 7, 60 °C) with continual molecular weight decrease in the first step [185]. The beginning of mass loss in the second step coincides with a decrease in the pH of the solution to a value of about 4. An autocatalytic effect of released hydroxy acids was confirmed by an increasing degradation rate. Moreover, a nonrandom chain scission with a more accelerated hydrolysis of the 3HV component was also found in these experiments. An inhomogeneous degradation with accelerated hydrolysis of the core was observed in thicker P3HB-7%3HV films [186]. The accelerated hydrolysis of the core, similar to that found in poly(lactide)s, was attributed to the accumulation of acidic degradation products in the polymer bulk.

In a study on P3HB-14%3HV fibers incubated in buffer solution (pH 7.4, 37 °C), the molecular weight M_n started to decrease after an induction period of about 80 days, and reached 64% of its initial value after 6 months. As confirmed in the accelerated test, the tensile strength and mass started to decrease at molecular weights M_n of about 70 000 and 17 000, respectively [187]. Textiles made from melt-extruded P3HB fibers showed a molecular weight decrease to about 90% of the initial values after 6 months of incubation (pH 7.4, 37 °C). The single fibers, however, degraded at a faster rate, reaching molecular weights M_w of 84% after 12 weeks in vitro, which could be further accelerated after pretreatment by electron irradiation [105]. P3HB "wool" made from solution-spun fibers have been tested under accelerated conditions (pH 10.6, 70 °C), confirming the influence of polymer purity and processing on the hydrolysis rate [188].

Porous scaffolds made from P3HB-3%3HV and P3HB-3%3HV/wollastonite composites showed a continuous fast decrease in their mass and molecular weight when stored for up to 15 weeks in buffer solution (pH 7.4, 37 °C). For example, unmodified P3HB-3HV scaffolds lost about 12% of their initial mass and about 90% of their initial molecular weight at the end of the incubation period. Increasing the amount of wollastonite in the composite resulted in increased mass loss due to wollastonite dissolution but delayed polymer hydrolysis, which was explained by the buffering effect of acidic hydrolysis products by alkaline ions dissolved from the wollastonite. This was confirmed by pH measurements showing a drop during the incubation of unmodified P3HB-3HV, but almost constant values when testing the composites [189]. Porous P3HB scaffolds tested in another study showed a mass loss of only 3% after 50 days of incubation in buffer solution (pH 7.4, 37 °C) [124]. Mass loss of P3HB-8%3HV porous scaffolds started after an induction period of 120 days and reached about 40% after 180 days in vitro (pH 7.4, 37 °C) [126, 127].

The degradation behavior of P3HB was also studied for the selection of suitable sterilization methods. It was found that with steam sterilization and with gamma-irradiation in particular a molecular weight reduction and deterioration of mechanical properties takes place, while sterilization with ethylene oxide or formaldehyde gas has no affect on the polymer properties [190]. Significant loss in the molecular weight together with embrittlement have been observed for P3HB films, but not P3HB-3HV films, in a screening study testing a variety of sterilization methods [102]. The irradiation-induced degradation of P3HB or P3HB-3HV was studied and confirmed also by other authors [105, 191–196]. However, it was also reported that P3HB-3HV can be sterilized by steam, gamma-irradiation, and ethylene oxide/carbon dioxide without losing stress and stiffness [197].

In Vivo Degradation

The in vivo degradation (decrease of molecular weight) or resorption (mass loss) of P3HB has long been a controversial subject in the literature. Main reasons for the controversy were the use of samples made by various processing technologies (e.g., solution-casting, melt-processing, drawing/orientation) in different shapes and designs (e.g., films, fibers, porous scaffolds) and the incomparability of different implantation sites (e.g., blood contact, "soft" tissue, "hard" tissue). However, there is now a significant body of research data available confirming that P3HB is a completely resorbable polymer, with a degradation rate comparable to that of slowly degrading synthetic polyesters such as high molecular weight PLLA. Most valuable for an estimate of the in vivo resorption time of P3HB are studies conducted in sheep which demonstrated the resorption of arterial implants (blood contact) after approximately 12–24 months and pericardial implants ("soft" tissue contact) after approximately 30 months (see below).

The following conflicting results reported from early studies may serve as examples for the controversy previously existing in the literature. For example, solution-cast and press-sintered P3HB showed beginning or complete dissolution after i.m. and s.c. implantation in rabbits already within 2–8 weeks [26]. P3HB tablets lost about 7.5% of their mass within 20 weeks s.c. in mice [28]. In contrast, no decomposition of P3HB tablets was observed after a 6-month period s.c. in rats [198]. No significant changes of P3HB films were reported from a 90-day s.c. implantation study in rats [199].

It has to be considered that mass loss or decomposition of P3HB occurs in the final step of its degradation, as already outlined above. Therefore, in comparison to mass loss, molecular weight analysis gives a better insight into the degradation profile of a bulk-degrading polymer such as P3HB and allows for a more reliable prediction of its resorption time.

For example, various P3HB samples implanted for 1 year s.c. in rats showed a decrease in the molecular weight to about 70–85% of the initial value [192]. Systematic degradation studies of polyesters, among them P3HB, were carried out s.c. in mice [55]. The 2% mass loss of P3HB after 6 months of implantation was ascribed to low molecular weight impurities. The molecular weight was reduced to 57% of the initial value. The in vivo degradation rate had the order PDLLA > PLLA > P3HB (Fig. 10a). It was concluded that P3HB degrades in vivo, although at a much slower rate than the poly(lactide)s.

Additionally, the in vivo degradation of P3HB was systematically compared with that of P3HB-3HV in this study (Fig. 10b). P3HB-3HV copolymers show a slower degradation than P3HB if the 3HV content is less than 10%. However, the hydrolysis rates become comparable when the 3HV content

a)

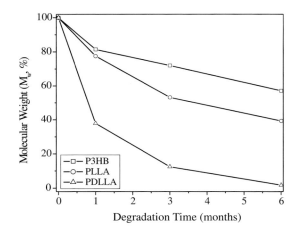

b)

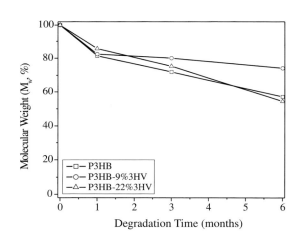

Fig. 10 **a** In vivo degradation (s.c., mouse) of injection-molded samples of P3HB, compared to that of PLLA and PDLLA, and **b** to that of P3HB-9%3HV and P3HB-22%3HV [55]

reaches about 20%. This might be explained by a slightly decreased polymer crystallinity with increasing amount of 3HV, which overshadows the decelerating effect caused by the increasing hydrophobicity. However, crystallinities of P3HB-3HV copolymers are generally high and in the order of magnitude of that of the P3HB homopolymer [65, 200] so that degradation rates can be expected that are similar for both types of polymers.

A slow polymer degradation has also been suggested from a long-term implantation study of P3HB-3HV (7%, 14%, 22% 3HV) i.m. in sheep. The copolymer containing 22% 3HV appeared to be more sensitive to the in vivo degradation than the copolymers having a lower 3HV content [100].

P3HB samples implanted s.c. and i.p. in rats showed a fast initial degradation in the first 4 weeks. The molecular weight had halved 1 year after implantation [201]. P3HB-7%3HV showed a decrease in the molecular weight to about 40% of the initial value 15 months after implantation s.c. into rabbits [184]. In a 4-week implantation study of P3HB-12.6%3HV films s.c. in rats it was shown that the polymer degradation can be accelerated by incorporation of a basic drug [182].

No mass loss of P3HB fibers was observed after s.c. implantation in rats for 180 days [202]. A slow resorption of P3HB fibers has also been reported after i.m and i.p. implantation in rats. The maximum breaking force was between 94% (i.m.) and 81% (i.p.) of the initial values 12 weeks postoperatively [203].

Nonwoven P3HB patches were implanted for closure of an atrial septal defect [144], as transannular patches for the enlargement of the right ventricular outflow tract and pulmonary artery [145] and as pericardial patches [142] in sheep. At 12 months postoperatively, no remaining polymer could be observed by light microscopy in the atrial septal defect study but small particles appeared in polarized light. The transannular patches disappeared microscopically after about 12 months; however, macrophages were still present 24 months postoperatively. The pericardial patches degraded more slowly. Some polymer particles remained 24 months after implantation, but disappeared after 30 months leaving behind some macrophages marking the implant region at this time. In clinical use, P3HB pericardial patches showed a significant reduction in size, by an average of 27%, in the 24-month follow-up group [33]. The complete absorption of a P3HB pericardial patch, 16 months after implantation in a patient, has also been reported [204].

Bioresorbable intravascular stents made of P3HB plasticized with TEC were nearly completely dissolved 16–26 weeks after implantation into the iliac arteries of rabbits [205, 206]. P3HB conduits for nerve regeneration showed softening [36] or fragmentation and size reduction [34] after a 12-month implantation in cats. In another study, the canine urethra was replaced by a polyglactin mesh coated with P3HB. The complete resorption of the graft was observed 8–12 months postoperatively [207].

Gastrointestinal patches made of P3HB/at-P3HB (70/30) blends tested for repair of a bowel defect in rats showed a fast initial degradation followed by a deceleration after 2 weeks [38]. Interestingly, this time period corresponds to the occurrence of mRNA encoding of pancreatic enzymes after implantation of P3HB patches onto the gastric wall of rats [150]. Material remnants were found only in one out of four test animals 26 weeks after implantation, and had a molecular weight of about 38% of the initial value (Fig. 11a) [38].

a)

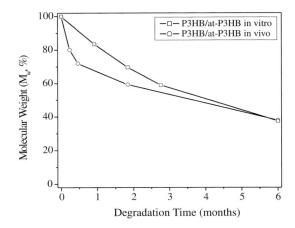

b)

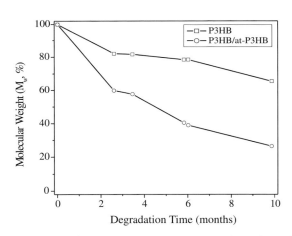

Fig. 11 a In vivo degradation of solution-cast P3HB/at-P3HB (70/30) patches implanted on the intestinal wall of rats [38]; in vitro degradation (pH 7.4, 37 °C) for comparison. **b** In vivo degradation of solution-cast P3HB and P3HB/at-P3HB (70/30) patches implanted to cover skull defects in minipigs [72]

P3HB films were tested subperiostally on the osseus skull of rabbits. Although there were no macroscopic signs of resorption after 20 months of implantation, no remaining test body could be found after 25 months [208]. Patches made of P3HB and P3HB/at-P3HB (70/30) blends have been used to cover skull defects in minipigs and a strongly enhanced degradation of the patch containing at-P3HB has been observed (Fig. 11b) [72]. An amount of 50% at-P3HB in the blend did not further accelerate the degradation of P3HB [39].

P3HB implants reinforced with carbon fiber for fixation of tibial osteotomies in rabbits were completely resorbed 24 weeks after implantation [209]. In contrast, P3HB used for bone regeneration of mandibular defects in rats did not show any signs of degradation in the 6-month study [210]. Examinations of P3HB/HA composites implanted in the femur of rabbits showed physical cracks after 3 months but no signs of resorption [93]. However, a degrading P3HB phase was observed at the interface between bone and pins made of P3HB or P3HB/HA after 6 months of implantation into condyles of rabbits [155]. P3HB screws implanted for 2 years intramedullary in sheep showed a slight reduction of hardness [153]. An almost completely resorbed osteosynthesis screw made of P3HB-15%3HV was found after 32 months of subfascial implantation in the sheep, whereas other implants, among them some made of P3HB homopolymer, showed no signs of resorption at this time. However, a large numeric difference between implanted and explanted samples was observed. Thus, in one sheep, despite a most thorough investigation, only 3 out of 15 implants could be retrieved [154].

3.4
Applications in Tissue Engineering

Cardiovascular Patches

P3HB patches have shown potential for guided tissue regeneration in a number of animal studies and in patients. Nonwoven P3HB patches made from solution-spun fibers have been extensively studied for defect repair in heart surgery, such as for pericardial substitution [32, 33, 142, 143, 204, 211–213], for closure of atrial septal defects [144], and for enlargement of the right ventricular outflow tract and the pulmonary artery [145].

Pericardial adhesions following open-heart surgery, which may severely complicate cardiac reoperations, were minimized and the coronary anatomy was preserved after implantation of a nonwoven P3HB patch in sheep for up to 30 months. The patch surface was shown to support regeneration of mesothelial cells with preserved prostacyclin activity [142, 211]. However, when tested after cardiopulmonary bypass surgery, which causes additional insult to the pericardium, implantation of P3HB pericardial patches did not affect adhesion formation or postoperative coronary anatomy visibility in a short-term study in sheep [212].

P3HB pericardial patches have been tested in a randomized clinical study and significantly limited retrosternal adhesions after cardiac surgery [33]. Another clinical report confirms the potential of P3HB pericardial patches for pediatric patients undergoing open heart surgery to repair congenital defects [204]. However, the patch does not affect postoperative changes in the right ventricular function after coronary artery bypass surgery [213].

Furthermore, nonwoven P3HB patches were effectively tested in sheep for cardiovascular repair in low pressure systems, such as for closure of an atrial septal defect [144] or as transannular patch in the right ventricular outflow tract [145]. There were no signs of aneurysm formation or calcification. Interestingly, no preseeding with vascular cells was necessary in these experiments because the high mechanical strength and slow degradation of the polymer prevented any dilatation of the regenerating tissue.

Gastrointestinal Patches

Resorbable P3HB patches for the gastrointestinal tract have been developed to cover large open lesions if closure by conventional surgical techniques with sutures or clips is impossible. These asymmetric patches were designed to have a porous surface facing the bowel defect to support tissue regeneration, and a smooth surface to prevent adhesions to the surrounding gut (Fig. 12).

a)

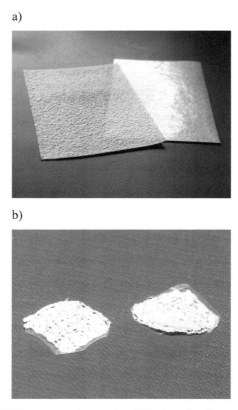

b)

Fig. 12 **a** P3HB/at-P3HB asymmetric patches [17] (reprinted by permission of Wiley) and **b** P3HB film/textile composite patches (Institute of Polymer Research, Dresden, Germany) with porous surface to support tissue regeneration and smooth surface to prevent adherence to surrounding organs

Tests have been made on patches made either from P3HB films [150], P3HB films blended with at-P3HB in order to reduce stiffness and enhance hydrolysis [38], or from P3HB film/textile composites [214]. A similar asymmetric patch, made either from P3HB/at-P3HB (70/30) or (50/50) blends, has recently been tested in minipigs and shown to support healing of a dural defect on its porous surface and prevent adhesions to brain tissue on its smooth surface [39].

Furthermore, P3HB tubes have been described that are intended to support tissue healing of vessels and hollow organs. P3HB was selected as a material that can resist the aggressive secretions of the intestinal tract for a sufficiently long time but will finally degrade [31]. It was demonstrated in a rat model that high-risk anastomoses under the condition of general peritonitis can be significantly protected by an intraluminal bypass with P3HB-7%3HV tubes. No postoperative complications due to adhesions and no signs of toxic reactions, as well as a good healing process with excellent revascularization of the anastomotic region have been observed [215].

Nerve Conduits

Conduits for peripheral nerve regeneration were formed from nonwoven P3HB sheets having polymer fibers oriented in one direction. These sheets were wrapped around a 2–3 mm gap of the transected superficial radial nerve in cats, and were sealed with fibrin glue [34, 36]. The fibers were oriented longitudinally in order to provide contact guidance and mechanical support for growing axons. The slow degradation of P3HB was considered to be of advantage, supporting the nerve during a period long enough for regeneration to take place and reducing the risk of accumulation of large amounts of acidic degradation products. Axonal regeneration was found to be similar to that observed with epineural repair, which was used as a control.

P3HB tubes with unidirectional fiber orientation were prepared by rolling a sheet of nonwoven P3HB around a cannula and sealing it with cyanoacrylate glue. These tubes were successfully used to bridge a 10 mm gap in the sciatic nerve of rats [35]. An increased axonal regeneration was found by filling the tubes with transplanted Schwann cells [216]. Allogenic Schwann cells, embedded in an alginate/fibronectin matrix inside the P3HB conduit, were shown to enhance axonal regeneration of the transected rat sciatic nerve without eliciting a deleterious immune response [217, 218]. Additionally, P3HB conduits filled with leukemia inhibitory factor in an alginate/fibronectin matrix enhanced the repair of the rat sciatic nerve in comparison to conduits without growth factor, but still did not perform as well as autografts [219].

Tubes made from P3HB sheets were also tested as conduits for long-gap peripheral nerve repair. P3HB tubes used to bridge up to 40 mm-long gaps in the transected peroneal nerve in rabbits supported nerve regeneration, although not as well as autografts. The authors concluded that empty P3HB conduits may not be optimal for long-gap repair, and improvements might be considered by using exogenous growth factors or cultured cells [220]. In a subsequent study using the same animal model, these tubes were filled with glial growth factor embedded in an alginate/fibronectin matrix. While cross-sectional staining of axons was greatest in the group containing growth factor, empty P3HB tubes showed the greatest axonal regeneration distance. Alginate had an inhibitory effect on the regeneration [221].

P3HB fibers embedded in an alginate/fibronectin matrix were tested as a scaffold in a spinal cord hemisection model in rats. Scaffolds, 2–3 mm long, supported neuronal survival after spinal cord injury. In the absence of polymer fibers, implants of alginate and/or fibronectin hydrogel alone were found to have no effect on neuronal survival [37].

Bone and Cartilage Repair
The long-term degradation profile of P3HB is considered to be of advantage in orthopedic applications [152, 222]. Thus for example, fast-degrading polyesters are not suitable for load-bearing fracture fixation devices [183], and bone implants made from those polymers have been coated with P3HB in order to slow down the degradation [223]. Additionally, the piezoelectric potential of P3HB has been considered as a special feature, because it is comparable with that of natural bone [224, 225]. It is well-known that bone can be strengthened and repaired by electrical stimulation. Therefore, P3HB composites may stimulate bone growth and healing [226].

P3HB implants have been tested for connecting osteotomies in the tibia of rabbits. The defect was healed after 12 weeks in most cases and the implant was completely resorbed after 24 weeks [209]. P3HB was also successfully tested as an occlusive membrane for guided bone regeneration in the mandibula of rats [210, 227]. Osteosynthesis plates made of P3HB and anchored with two P3HB bolts were examined for repair of a cut through zygomatic arches of rabbits. The polymer was assessed to be suitable for covering defects of the osseus skull and as an osteosynthesis material for fractures of the visceral cranium [208]. However, no differences appeared in the healing pattern of 15 mm diameter rhinobasal skull defects in minipigs when covered with a 20 mm diameter P3HB/at-P3HB patch, compared to the control group without patch [39].

P3HB composites were extensively examined for bone repair, such as composites made of P3HB and HA. HA stimulates the formation of new bone, while P3HB is considered as potentially bioactive due to its degradability and piezoelectricity. P3HB/HA composites were found to closely match the mechanical properties of cortical bone, but had insufficient strength and ductility for the construction of major load-bearing components [93]. Phosphate glass was added to improve mechanical strength and to serve as precursor for bone-forming osteoblasts [152]. Among composites made from P3HB or P3HB-3HV (8%, 12%, 24% 3HV) and HA, the P3HB-8%3HV/HA (30/70) composite had a compressive strength in the order of that of human bone, making it a candidate for fracture fixation [228]. Composites for cortico-cancellous bone grafts were also made from P3HB-7%3HV and HA [229].

Composites made from P3HB-7%3HV and TCP showed a good biocompatibility in the femur of rabbits. However, PLLA/TCP was regarded as more suitable for fracture support due to accelerated degradation, thereby maintaining the required strength for a bone healing time of about 6 weeks [184]. In vitro studies have shown the formation of bone-like apatite on the surface of P3HB/HA [230] and P3HB-12%HV/TCP [231] composites. Recently, the fabrication and properties of P3HB-3%3HV/wollastonite composite scaffolds have been reported [189, 232].

In vitro studies using bone marrow cells have shown the potential of P3HB and P3HB-8%3HV matrices for bone tissue engineering (see Sect. 3.2). A chemically synthesized degradable polyesterurethane foam containing crystalline domains of short-chain P3HB and amorphous domains of PCL exhibits good osteoblast compatibility and is being considered as scaffold material for bone repair [233].

The potential of P3HB films and scaffolds as matrices for cartilage regeneration have been extensively studied in vitro (see Sects. 3.2 and 5.2). Thus for example, nanofibrous polymer mats have been fabricated by electrospinning of P3HB-5%3HV solution and tested as potential scaffolds for chondrocyte attachment [133]. However, animal studies have not yet been reported.

Other Tissue Engineering Applications

P3HB films [234] and P3HB-coated textiles [235] have been described as materials for wound covering. Nonwoven P3HB "wool" has been developed as an internal wound scaffolding device that supports reepithelialization [236].

P3HB surface-modified by ammonia plasma has been examined as a replacement material for respiratory mucosa after surgical resections in the trachea, nasal septum, or sinus maxillaris [136]. P3HB-8%3HV modified by oxygen plasma has been tested as a temporary substrate for retinal pigment epithelium cells, to treat retinal disorders caused by retinal pigment epithelium degeneration [137].

4
Poly(4-hydroxybutyrate)
and Poly(3-hydroxybutyrate-*co*-4-hydroxybutyrate)

P4HB is a thermoplastic polyester that can be produced by a number of microorganisms, including *Alcaligenes eutrophus* strains [237–239], *Comamonas acidovorans* [240], *Hydrogenophaga pseudoflava* [241], and recombinant *Escherichia coli* [47, 242, 243], using substrates such as 4-hydroxybutyrate, 4-hydroxybutyrolactone, or 1,4-butanediol. P3HB-4HB can be obtained by fermentation using microorganisms such as *Alcaligenes eutro-*

phus [244], *Pseudomonas acidovorans* [245], *Comamonas acidovorans* [246], or *Hydrogenophaga pseudoflava* [241]. Biotechnologically produced P4HB and P3HB-4HB have high molecular weights and properties suitable for medical applications, including tissue engineering (congenital heart defects, heart valves, vascular grafts), suture materials, and surgical textiles [47].

4.1
Mechanical Properties

P4HB has improved mechanical properties over P3HB, such as low stiffness and brittleness, and high elongation at break. Furthermore, improved properties can already be found in P3HB-4HB copolymers having small amounts of 4HB [62]. With increasing amounts of 4HB, to approximately 20–50%, these copolymers exhibit elastomeric properties (Martin DP, personal communication). Thermal and mechanical properties of P4HB and P3HB-4HB solution-cast films are compared with those made of P3HB in Table 2 [51, 247].

Slightly different mechanical data for P4HB are reported from another study, comparing dense and porous films for cardiovascular tissue engineering made by melt-processing without or with salt-leaching. The mechanical strength of porous scaffolds is low compared to dense polymer films (Table 3) [56].

Table 2 Thermal and mechanical properties of P3HB, P3HB-4HB, and P4HB [51, 247]

Polymer	T_g	T_m	Elastic modulus	Tensile strength	Elongation at break
	[°C]	[°C]	[MPa]	[MPa]	[%]
P3HB	10	178	2730	36	2
P3HB-18%4HB	– 22	137	320	12	1120
P4HB	– 47	61	230	36	1140

Table 3 Mechanical properties of dense P4HB films (made by melt-processing) and porous P4HB films (made by melt-processing/salt-leaching) [56]

P4HB morphology	Elastic modulus [MPa]	Tensile strength [MPa]	Elongation at break [%]
Dense film	64.8	51.7	1000
Porous film (50% porosity)	14.9	6.2	164
Porous film (80% porosity)	1.8	1.2	100

4.2
Biocompatibility

Toxicity Testing
P4HB and P3HB-4HB have been evaluated in preclinical tests recommended by the FDA for medical devices. These tests include cytotoxicity, sensitization, irritation and intracutaneous reactivity, hemocompatibility, and implantation. Thus for example, P4HB films and sutures were subjected to a complete series of biocompatibility test protocols that were performed in accordance with the FDA's GLP regulations as set forth in 21 CFR, part 58, as well as ISO 10993-1. The test results confirmed that P4HB is nontoxic and biocompatible (Martin DP, personal communication).

The degradation product 4HB is a natural metabolite present in the human brain, heart, lung, liver, kidney, and muscle [248]. It has recently been approved by the FDA for treatment of cataplexy [249]. In higher doses, 4HB possesses psychopharmaceutical effects leading to abuse as "liquid ecstasy" [250]. However, the low amount of monomer released during the polymer degradation together with the short in vivo half-life of 4HB makes it highly unlikely that small P4HB implants, such as those used in cardiovascular tissue engineering, will cause pharmacological side-effects [47].

Cell Culture Studies
In vitro cytotoxicity studies using L929 mouse fibroblasts revealed comparable cell compatibility of P3HB, P4HB, and P3HB-18%4HB films [51]. Other cell types have been successfully seeded on P4HB and tested in tissue engineering experiments, mostly for cardiovascular applications. For example, porous P4HB matrices have been seeded with ovine vascular cells, and cell attachment and growth was compared to that on PGA or P3HO-3HH matrices. Although cell attachment on P4HB was low, it was considered to be sufficient for the application as heart valve scaffold [44]. Porous P4HB scaffolds seeded with ovine vascular cells have also been tested as a cardiovascular patch [42] or vascular graft [46, 251]. PGA matrices coated by a P4HB film layer have been successfully seeded with different types of cells, such as ovine vascular myofibroblasts and endothelial cells [43, 45, 252, 253], endothelial progenitor cells [254], marrow stromal cells [255–257], human pediatric aortic cells [258], human umbilical cord cells [259], human mesenchymal placental cells [260], or ovine skeletal myoblasts [261] (see Sect. 4.4).

In Vivo Studies
Mild tissue reactions have been observed in several in vivo studies. For example, a minimal foreign body reaction has been reported after s.c. implantation of P4HB in rats [56]. The analysis of patches tested to augment the pulmonary artery in sheep revealed a little remaining polymer 24 weeks post-

operatively surrounded by fibrous tissue and eliciting a moderate mononuclear and giant cell reaction, which was limited to the polymer itself without affecting the surrounding tissue [42]. No signs of inflammation such as leukocyte or giant cell infiltration were observed when porous P4HB tubes were implanted as substitutes of the aorta in sheep [262]. No evidence of inflammation or residual polymer have been found 16–20 weeks after implantation of PGA/P4HB composite heart valve scaffolds in sheep [43].

Remnants of P4HB films implanted as aortic tissue substitutes in rabbits were surrounded by macrophages and a few lymphocytes 26 weeks after implantation. The inflammatory response to the degrading polymer was mild to moderate depending on the progress of degradation (see Sect. 4.3). It was assessed to be adequate by considering the final stage of polymer degradation associated with significant release of degradation products. There was no biomaterial encapsulation. P4HB films supported neotissue formation, which consisted lumenally of myofibroblasts, smooth muscle cells, and a layer of endothelial cells which was, however, partly incomplete. No excessive intimal proliferation and lumen narrowing, and no calcifications have been observed with these patches. P3HB-4HB patches tested in a similar model were surrounded by very few macrophages and there was no inflammatory cell infiltration due to the slow degradation process, without histologic signs of mass loss at 26 weeks. The polymer film was encapsulated. A moderate to strong intimal proliferation associated with slight lumen narrowing but no calcifications has been observed in this group. The endolumenal endothelialization was partly incomplete [263].

Rods made of P3HB-4HB implanted in the tibia of rabbits exhibited a good compatibility without adverse tissue reaction after 6 weeks [264].

Blood Compatibility
An activation of the cellular coagulation similar to that induced by P3HB was demonstrated after contact of P4HB and P3HB-4HB films with fresh human blood. An initial activation of the plasmatic coagulation was observed with P3HB-4HB films. However, both polymers did not activate the complement system, and the amounts of C3a-desArg released have been lower than those found with P3HB in these in vitro experiments (Fig. 13) [265].

In vivo, there was no thrombus formation despite partly incomplete endothelialization on either P4HB or P3HB-4HB films tested as patches in the rabbit aorta [263]. No signs of thrombus deposition were reported from studies on tissue-engineered PGA/P4HB heart valves [43] and P4HB pulmonary artery patches [42] implanted in sheep. However, tissue-engineered P4HB tubes replacing the aorta in sheep induced thrombus formation, and increasing deposits have been detected between 12 and 24 weeks postoperatively [262].

It is worth noting that the degradation product 4HB reduces the human platelet aggregation induced by collagen and arachidonic acid, the latter even

a)   b)

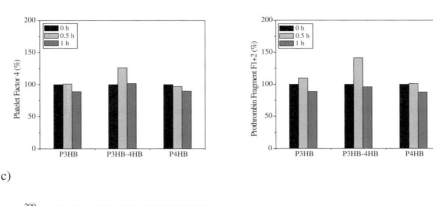

c)

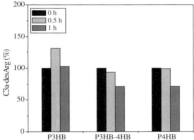

Fig. 13   In vitro hemocompatibility of solution-cast P3HB, P3HB-18%4HB, and P4HB films. Activation of **a** cellular coagulation (release of platelet factor 4), **b** plasmatic co-agulation (release of prothrombin fragment F1+2), and **c** complement system (release of C3a-desArg) [265]

more potently than aspirin. Similarly, the thrombin-triggered aggregation is reduced and the thromboxan production is decreased [266].

4.3
Biodegradability

In Vitro Degradation
An accelerated degradation rate of P4HB and P3HB-4HB, in comparison to P3HB, can be expected due to the lower degree of crystallinity. This was con-firmed by studying molecular weight decrease and mass loss of polymer films incubated in buffer solution (Fig. 14) [51]. Other studies demonstrated a mo-lecular weight decrease of about one third within 10 weeks of incubation in buffer solution (37 °C, pH 7.4) [56]. Even small amounts of 4HB in P3HB-4HB copolymers lead to a significantly faster hydrolysis compared to that of P3HB. Increasing the amount of 4HB in the copolymers resulted in an acceler-

a)

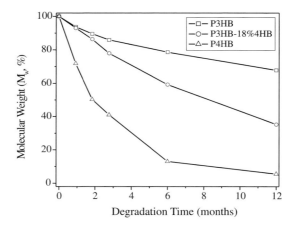

b)

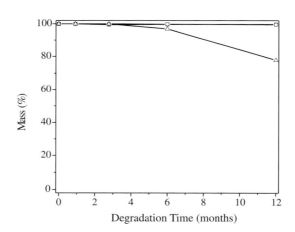

Fig. 14 In vitro degradation (pH 7.4, 37 °C) of solution-cast P3HB, P3HB-18%4HB, and P4HB films. **a** Molecular weight and **b** sample mass vs. degradation time [51]

ated hydrolysis in these experiments [167, 267], and the molecular weights M_n of P3HB-9%4HB and P3HB-16%4HB films reached approximately 60% and 40%, respectively, after 160 days in buffer solution (37 °C, pH 7.4), compared to 80% for the P3HB homopolymer [167].

In Vivo Degradation
The complete resorption of P4HB within 8 weeks has been reported when used as coating layer for a PGA heart valve scaffold in sheep [43]. A porous

P4HB patch for the augmentation of the pulmonary artery in sheep was nearly completely resorbed within 24 weeks [42]. A similar result has been reported with porous P4HB tubes replacing a part of the descending aorta in sheep [262]. Compact P4HB films implanted s.c. in rats showed a molecular weight decrease to approximately 62% after 10 weeks while highly porous P4HB samples were completely resorbed at that time. Films, 50%, and 80% porous samples showed a 20%, 50%, and 100% mass loss, respectively, over the 10 week period. Microscopic examination revealed a progressive surface erosion starting with signs of cracking after 1 week. On the other hand, in vitro samples showed almost no discernable change during 10 weeks of incubation in buffer solution, suggesting a biologically mediated mode of degradation in vivo [56].

P4HB films tested as aortic patches in rabbits showed progressive signs of degradation, confirming a surface erosion mechanism (Fig. 15); films made of

a)

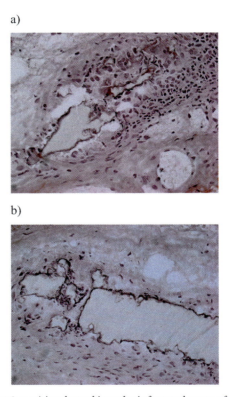

b)

Fig. 15 P4HB films (6 × 2 mm) implanted into the infrarenal aorta of rabbits 26 weeks postoperatively (H&E, 400×). **a** Nearly complete film degradation in one animal. **b** Surface erosion and beginning film fragmentation in another animal. The inflammatory response is focal and appropriate at this stage of polymer resorption, with the number of macrophages surrounding the film remnants dependant on the degradation progress [263]

P3HB-18%4HB were essentially unchanged 26 weeks after implantation [263]. Summarizing the in vivo observations reported thus far it can be concluded that P4HB has a resorption time of approximately 6–12 months, depending on the sample morphology and implantation site. This corresponds approximately to the degradation time of PDLLA in vivo.

P3HB-4HB copolymers can be expected to degrade with a rate between that of P4HB and P3HB. Thus for example, P3HB-10%4HB films have been reported to degrade to 80% of the initial molecular weight (M_n) when implanted for 4 months i.p. in rats [22] and P3HB-20%4HB films degraded to 55% of their initial M_n after 1 year of s.c. implantation in rats [192].

4.4
Applications in Tissue Engineering

P4HB and P3HB-4HB have primarily been tested as scaffolds in cardiovascular tissue engineering. Preliminary studies using heart valve scaffolds based on PGA or P3HO-3HH showed limitations. While PGA has insufficient mechanical properties, P3HO-HH is resorbed too slowly [268] (see Sect. 6.4). P4HB has been introduced as an alternative biomaterial with sufficient flexibility and suitable degradation time for applications in the circulatory system [47], and a large number of studies report on its potential for tissue engineering of heart valves and small-caliber vascular grafts, as well as its application as a patch material.

Heart Valves
Polymer heart valves made entirely from P4HB have been described, either constructed from highly porous polymer films (Fig. 16a) [44] or made by rapid prototyping. The latter allows for a resemblance of the human anatomy, as derived from computed tomography of patients [269]. However, the majority of studies on P4HB-based heart valve scaffolds have used a PGA/P4HB composite material made by dip-coating of a PGA mesh in P4HB solution. Trileaflet heart valves fabricated from the PGA/P4HB composite have been seeded with ovine carotid artery medial cells for 4 days under static conditions, followed by addition of endothelial cells and culture in a dynamic pulse duplicator for 14 days, and were implanted in sheep for up to 20 weeks. Echocardiography demonstrated functioning leaflets without stenosis, thrombus, or aneurysm. Complete degradation of the PGA mesh was observed after 4 weeks, and that of the P4HB layer after 8 weeks. Despite this fast scaffold degradation, mechanical properties of the regenerated valve leaflet tissue were comparable to those of native tissue at 20 weeks [43]. The remodeling process was followed by analysis of the cell phenotype of the tissue-engineered construct after in vitro and in vivo testing. Cells from in vitro constructs (14 days) were found to be activated myofibroblasts with strong expression of α-actin and vimentin; cells from in vivo explants

(16–20 weeks) were fibroblast-like, with expression of vimentin but undetectable levels of α-actin, which is similar to the native valve. The ECM architecture of the explanted tissue resembled that of native valves. However, leaflets were only partially covered with endothelial cells [252].

Other cell types investigated for preseeding of PGA/P4HB heart valve scaffolds include marrow stromal cells (MSCs), which exhibit the potential to differentiate into multiple cell-lineages and can be easily obtained clinically [255–257]. Ovine aortic valvular endothelial cells and circulating endothelial progenitor cells seeded on PGA/P4HB proliferate in response to vascular endothelial growth factor and transdifferentiate to a mesenchymal phenotype in response to transforming growth factor β1, which is a critical step during embryonic development of cardiac valves [254]. Furthermore, human pediatric aortic cells and factors influencing their maturation have been studied using PGA/P4HB scaffolds [258].

A number of studies based on PGA/P4HB have confirmed the importance of dynamic cell culture conditions that reproduce physiological mechanical stress and of pulsatile flow to achieve optimally functional scaffolds [43, 45, 46, 255, 259]. A bioreactor that includes dynamic flexure as a major mode of deformation in the native heart valve cusp has been tested on PGA/P4HB and PLLA/P4HB composite scaffolds [270].

A combination of an acellular porcine heart valve coated with P3HB, P4HB, or P3HB-4HB has recently been developed (Fig. 16b) [51]. The decellularized heart valve resembles the natural 3D structure of the heart valve, providing nearly ideal flow conditions. It is composed mainly of collagen, elastin, and proteoglycans, and contains protein ligands for cell attachment and receptor activation. The protein/polyester combination has been developed to overcome limitations of protein and polymer valves alone. Polymer coating of a tissue-derived acellular scaffold can improve the mechanical stability and

a)   b)

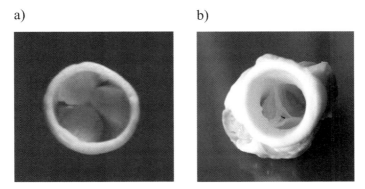

Fig. 16 **a** Trileaflet heart valve scaffold fabricated from porous P4HB [44] (reprinted by permission of Lippincott Williams & Wilkins). **b** Decellularized porcine heart valve as 3D matrix for P3HB, P4HB, or P3HB-4HB coating [51]

enhance the hemocompatibility of the protein matrix. This approach also reduces the amount of polymer used, facilitating degradation and increasing the elasticity of the construct [51]. Functional testing of these composite valves have been performed under physiological systemic load conditions using a pulse duplicator system [247, 271]. Decellularized porcine heart valves coated with P3HB-4HB have been tested without any cellular preseeding in the aortic and pulmonary position in sheep. While the two valves in aortic position performed well, with complete endothelialization and limited inflammatory cell invasion after 12 weeks, one of the two valves in the pulmonary position failed, probably due to bacterial endocarditis [272].

Vascular Grafts

PGA/P4HB dip-coated composites have been formed into tubular scaffolds using a heat application welding technique to test their potential for tissue engineering of blood vessels. The tubes were seeded with ovine vascular myofibroblasts for 4 days under static culture conditions, followed by seeding of endothelial cells and incubation for 28 days in a pulse duplicator system. The result was significantly more cells and collagen production as well as higher mechanical strength compared to the static controls [45]. Dynamic rotational seeding and culturing in a hybridization oven has also been shown to be an effective method to culture ovine vascular myofibroblasts onto these PGA/P4HB scaffolds [253].

P4HB porous tubes have been fabricated by a solution-casting/salt-leaching method in a cylindrical mold containing a cylindrical core. These tubular scaffolds (inner diameter 15 mm, wall thickness 2 mm) were seeded with ovine vascular smooth muscle cells, derived by enzymatic dispersion in order to limit the amount of differentiated myofibroblast-like cells. They were incubated for 4 days under static conditions and 14 days under dynamic conditions, followed by another static culture with endothelial cells for 2 days. The result was the creation of tissue-engineered aortic blood vessels. The dynamically cultured tubes showed confluent layered tissue formation with significantly increased ECM synthesis, DNA, and protein content compared to static controls, as well as mechanical properties approaching those of native aorta [46]. Protein expression profiles of the tissue-engineered P4HB tubes revealed distinct differences from native aorta or carotid arteries at this early stage of remodeling [251]. The tissue-engineered constructs were tested under high-pressure conditions in sheep by replacing a 4-cm segment of the descending aorta. The tubes were wrapped into decellularized ovine small intestinal submucosa prior to implantation in order to stimulate angiogenesis and increase mechanical stability. No intimal thickening, dilatation or stenosis were observed 12 weeks after implantation; however, small areas of thrombi were formed despite endoluminal endothelial cell-layering. Dilatation causing increased thrombus formation was observed 24 weeks after implantation (Fig. 17). Compared to native aorta, regenerated tissue con-

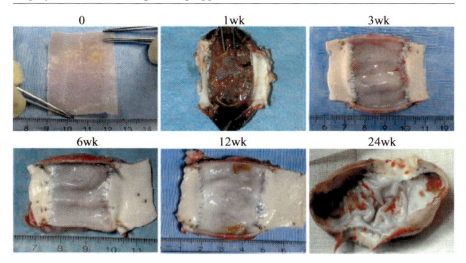

Fig. 17 Macroscopical appearance of tissue-engineered blood vessel grafts prior to implantation and 1, 3, 6, 12 and 24 weeks after implantation [262] (reprinted by permission of Elsevier)

tained significantly less elastin. The lack of mature, cross-linked elastic fiber formation has been identified as a major limitation of current tissue engineering approaches in the high-pressure circulation [262].

Cardiovascular Patches

Patch materials for augmentation of the pulmonary artery or the right ventricular outflow tract necessary in congenital cardiac defect surgeries have been made of highly porous P4HB films. In contrast to nonseeded P3HB-based cardiovascular patches (see Sect. 3.4), P4HB patches were seeded with autologous vascular cells prior to implantation into the pulmonary artery in sheep. Postoperative echocardiography of the seeded patches demonstrated a smooth surface without dilatation or stenosis. Histologically, formation of organized and functional tissue could be demonstrated 24 weeks after implantation. Despite an ingrowth of tissue from the surrounding native pulmonary artery onto the unseeded control patch, it showed a slight bulging, potentially indicating a beginning dilatation, after 20 weeks [42]. No aneurysm formation was reported from a similar study using unseeded P3HB patches [145]. Thus, to prevent aneurysm formation, cellular preseeding of biodegradable patches might be necessary, depending on the degradation time of the scaffold and the ability to provide mechanical strength for a sufficient period of time [42].

Tubular scaffolds made from PGA/P4HB composites have also been considered as vascular conduits to repair congenital cardiac defects such as surgical reconstruction of the right ventricle to pulmonary artery continuity. These scaffolds were seeded with human umbilical cord cells under static and dynamic culture conditions. Viable, confluently layered tissue with myofibroblast-like elements was formed after 21 days of culture. However, lack of elastin production was reported, causing low elongation at break and high stiffness of the scaffold compared to native tissue [259].

Other Tissue Engineering Applications

Fetal tissue engineering with constructs made of PGA/P4HB seeded with human mesenchymal placental cells has been tested as a potential method for treatment of severe congenital anomalies. P4HB was used to improve the mechanical strength of the PGA mesh in this study [260]. PGA/P4HB composites surface-treated with alkaline solution have been cultured with ovine skeletal myoblasts to test their potential for treatment of congenital muscular defects. However, they showed limited cell attachment compared to PGA/PLLA or collagen scaffolds [261].

5
Poly(3-hydroxybutyrate-*co*-3-hydroxyhexanoate)

P3HB-3HH is a thermoplastic polyester, which has been found to accumulate in *Aeromonas* species such as *A. caviae* [63] and *A. hydrophila* [273, 274]. Recently, the large-scale production of P3HB-3HH has been reported by fed-batch culture of *A. hydrophila* using glucose and lauric acid as carbon source. The resulting random copolymer of this fermentation process was found to have a 3HH fraction of 11% [275].

5.1
Mechanical Properties

Copolymerization of P3HB by incorporation of a longer alkyl side-chain such as 3HH is a promising strategy for improving the mechanical properties of the P3HB homopolymer because the 3HH unit does not fit into the crystalline lattice of 3HB, and vice versa, which avoids the isodimorphism found in P3HB-3HV copolymers [276]. The mechanical properties of P3HB-3HH are characterized by a low elastic modulus, high elongation at break, and relatively low tensile strength [63, 89, 90, 274] (Table 4). The mechanical strength of P3HB-3HH films can be improved by preorientation using cold-drawing [276] or hot-drawing [277] techniques.

Additionally, P3HB-3HH was shown to be a good candidate for blending in order to reduce the stiffness and brittleness of highly crystalline polymers such as P3HB [89, 90] and PLLA [278]. Improved mechanical properties have been found in P3HB/P3HB-13%3HH 40/60 blends, which were selected for subsequent biocompatibility studies [90]. PLLA/P3HB-5%3HH

Table 4 Mechanical properties of P3HB-3HH [274]

Polymer	Elastic modulus [MPa]	Tensile strength [MPa]	Elongation at break [%]
P3HB-2.5%3HH	630	25.7	7
P3HB-7.0%3HH	290	17.3	24
P3HB-9.5%3HH	155	8.8	43

and PLLA/P3HB-13%HH blends showed favorable mechanical properties up to a P3HB-3HH content of 20% [278].

5.2
Biocompatibility

Cell Culture Studies
The cell compatibility of P3HB-3HH and its blends with P3HB have been tested in a number of in vitro studies using fibroblasts [89, 94, 115–117], bone marrow cells [49, 94, 124, 279], and chondrocytes [48, 89, 90, 129–131]. However, neither in vivo testing nor hemocompatibility testing have yet been reported.

Published results with mouse fibroblasts seeded on polymer films are partially controversial. While a high cell viability on P3HB-3HH films and virtually no viability on P3HB films have been reported from earlier studies [89, 115, 116], comparable fibroblast viabilities on both materials have been described more recently [94]. Methods studied for surface modification to improve the wettability of the hydrophobic films include ion implantation [110], treatment with lipases [89, 115, 116], and alkaline hydrolysis [115, 116]. Surface hydrolysis using lipases or alkaline medium led to strongly improved cell viability on P3HB and P3HB/P3HB-3HH films while P3HB-3HH films remained essentially unaffected. Surfaces having higher hydrophilicity and smoothness were found to increase fibroblast viability in these studies [89, 115, 116]. The effect of the comonomer composition on the cell compatibility showed preferable growth on PHB-20%3HH in comparison to P3HB-5%3HH and P3HB-12%3HH surfaces, which was attributed to the surface smoothness. As already mentioned, no significant differences between P3HB and P3HB-3HH were found in this study. PLLA showed less cell compatibility than the PHB-based materials [94]. Comparable fibroblast viabilities have also been found after seeding on P3HB and P3HB-3HH porous matrices, made by solution-casting/salt-leaching and treated with lipase. Coating with hyaluronic acid had a negative effect on the cell growth in these experiments despite providing a smoother and more hydrophilic surface. It was concluded that surface properties have to be appropriate for protein adhesion and cell attachment [117].

Studies with rabbit bone marrow stromal cells have shown that these cells can attach, proliferate, and differentiate into osteoblasts when seeded on P3HB-11%3HH films. However, no phosphate deposits could be detected in the 28-day study [279]. A comparison of cell viabilities on P3HB, P3HB-3HH and PLLA films had the order: P3HB-12%3HH > P3HB-5%3HH > P3HB > P3HB-20%3HH > PLLA. This order is different to that found for fibroblasts and was attributed to the appropriate surface roughness promoting attachment and differentiation of osteoblasts-like cells, while fibroblasts prefer smoother surfaces [94]. Porous scaffolds made by solution-casting/salt-leaching confirmed the good cell compatibility of P3HB-3HH matrices, which showed higher cell viabilities and osteoblast differentiation (as determined by alkaline phosphatase activity) than scaffolds made from P3HB and PLLA. Additionally, a significant mineralization has been observed in this study [49]. Addition of HA to P3HB-3HH scaffolds resulted in reduced osteoblast growth, while the growth on P3HB/HA scaffolds was enhanced in comparison to the unmodified materials [124].

Rabbit articular cartilage-derived chondrocytes have been seeded on solution-cast films made from P3HB, P3HB-12%3HH, and their blends. Blends containing equal amounts of both polymers, possessing the highest surface free energy of the samples tested, had the highest amount of protein adsorbed and number of chondrocytes adhered [129]. Studies on porous scaffolds showed more chondrocyte growth on P3HB/P3HB-3HH blends than on the individual polymers alone within the 28-day culture period [48, 89]. Microscopic analysis revealed that large quantities of chondrocytes grew initially on the surface and, after 7 days, also into the open pores of the polymer scaffolds. Morphologically, cells found on the surface of the scaffold exhibited a fibroblast-like appearance and slowly formed confluent cell multilayers starting from 14–28 days of incubation. In contrast, proliferating chondrocytes, which maintain their morphology and phenotype for up to 28 days, have been found inside the polymer matrices [48]. High amounts of glycosaminoglycans and ECM (collagen) were produced in scaffolds made from the polymer blends. The level of collagen II that is secreted by maturating chondrocytes as cartilage-specific ECM protein was higher in P3HB-3HH or P3HB/P3HB-3HH than in P3HB scaffolds [130, 131]. Analysis of the ECM formed during cell culture with a scaffold containing 40% P3HB and 60% P3HB-3HH demonstrated high levels of calcium and phosphorus in a molar ratio of $Ca/P = 1.66$, which is similar to that of natural HA ($Ca/P = 1.67$), the major inorganic component of bone and cartilage [90].

5.3
Biodegradability

The (nonenzymatic) degradation of P3HB-3HH has not yet been tested systematically either in vitro or in vivo. Data has been reported from a short-term in vitro study of P3HB-12%3HV porous scaffolds showing 7% mass loss after 50 days of incubation in simulated body fluid (pH 7.4, 37 °C), in comparison to 3% for P3HB [124]. Two opposite effects on the hydrolysis process have to be considered when comparing the degradability of P3HB-3HH with that of the P3HB homopolymer. Firstly, the introduction of 3HH side-chains into P3HB increases the hydrophobicity of the polymer, thus hindering the penetration of water into the polymer bulk and slowing down the hydrolysis. Secondly, 3HH groups decrease the polymer crystallinity [63], which would result in an accelerated hydrolysis compensating the effect of the more hydrophobic side-chains. Overall, it can be assumed that the degradation rate of P3HB-3HH is comparable or slightly faster than that of P3HB or P3HB-3HV.

5.4
Applications in Tissue Engineering

Bone and Cartilage Repair

Based on the results of the in vitro cell compatibility testing (see Sect. 5.2), which showed improved attachment, proliferation and differentiation of rabbit bone marrow cells on P3HB-3HH films [94, 279] and porous matrices [49, 124] in comparison to P3HB or PLLA, it was concluded that P3HB-3HH has potential as scaffold material in bone tissue engineering. However, there is still a lack of in vivo data. In vitro, a strong alkaline phosphatase activity as an early marker of osteoblast differentiation has been reported; however, phosphate deposits as a late marker of osteoblast differentiation could be detected only on porous materials but not on dense polymer films after 28 days of cell culture [49, 279]. Composite scaffolds made from P3HB-3HH and HA as a bioactive and osteoconductive additive did not enhance osteoblast growth in vitro [124]. Ultimately, P3HB-3HH matrices have to be tested in vivo to confirm their potential for bone repair.

Similarly, despite a number of in vitro studies demonstrating the potential of P3HB-3HH films [129] and porous matrices [48, 89, 90, 130, 131] for engineering of cartilage (see Sect. 5.2) no in vivo results have been reported so far. In vitro, blends made from P3HB and P3HB-3HH showed most promise, in comparison to P3HB and P3HB-3HH alone, as they enhanced not only adhesion and proliferation of rabbit chondrocytes but also allowed for preservation of the cell phenotype, which is required for chondrogenesis [48]. Additionally, these materials allow for synthesis of hyaline cartilage-specific collagen II, which is the major component of the solid matrix of human articular cartilage [130, 131]. Another major ECM protein secreted by chondrocytes at the late stage of cartilage development, collagen X, could also be detected [129]. These important indicators of functional tissue development observed in vitro have to be confirmed in vivo to assess the potential of P3HB-3HH in cartilage engineering.

6
Poly(3-hydroxyoctanoate-*co*-3-hydroxyhexanoate)

P3HO-3HH has been biosynthesized in a fed-batch fermentation process using *Pseudomonas oleovorans*, with sodium octanoate as the sole carbon source. In this process, the metabolism of the bacteria can either add or cleave off two carbons from the fed octanoate molecule and produces not a homopolymer but a random copolymer, which has a typical composition of 86% 3-hydroxyoctanoate (3HO), 11% 3-hydroxyhexanoate (3HH), and 3% 3-hydroxydecanoate (3HD) [58]. Because of the major component, the copolymer is often simply referred to as poly(3-hydroxyoctanoate), P3HO. The production of medical grade P3HO-3HH, its properties and modification, as well as the fabrication of scaffolds for engineering of vascular grafts and heart valves have been reviewed [3].

6.1
Mechanical Properties

P3HO-3HH is a thermoplastic elastomer with a low tensile modulus, high elongation at break, and high tensile strength due to the orientation of the amorphous rubbery chains [52, 58]. As for other semicrystalline PHAs, the mechanical properties of P3HO-3HH are affected by the thermal history, such as crystallization temperature and time for crystallization/annealing [58]. Furthermore, the mechanical properties of P3HO-3HH are dependent on the morphology of the sample specimens, which is important for tissue engineering applications. For example, highly porous films made by a combination of solution casting and salt leaching techniques [280] have a significantly lower modulus and strength compared to dense films made by simple solution casting [52] (Table 5). Porous films have mechanical properties that are in the range of those of natural cardiovascular tissue [280].

Table 5 Mechanical properties of dense P3HO-3HH films (made by solution-casting) [52] and porous P3HO-3HH films (made by solution-casting/salt-leaching) [280]

P3HO-3HH morphology	Elastic modulus [MPa]	Tensile strength [MPa]	Elongation at break [%]
Dense film	10.4	45.1	198
Porous film	0.7	0.7	161

6.2
Biocompatibility

Cell Culture Studies
Cell culture experiments using a mixed population of ovine vascular cells (consisting of smooth muscle cells, endothelial cells, and fibroblasts) showed a significantly lower number of cells adherent to P3HO-3HH porous films as compared to PGA mesh, which was explained by the different porosities (50–60% for P3HO-3HH vs. 95% for PGA). P3HO-3HH and P4HB porous matrices showed comparable cell numbers in these experiments. Additionally, the deposition of ECM (collagen) on P3HO-3HH was low in comparison to that on PGA and P4HB [44]. There was no detectable elastin deposition in similar experiments [281].

In Vivo Studies
The biocompatibility of P3HO-3HH microspheres, tubes, and pellets have been tested by s.c. implantation in mice for up to 40 weeks. A very mild host reaction to the implants with formation of a thin fibrotic layer encapsulating the materials was described [3]. A good biocompatibility of P3HO-3HH have also been demonstrated in rats after s.c. implantation of P3HO-3HH-impregnated Dacron prostheses. The secretion of alkaline phosphatase by polymorphonuclear cells (acute inflammatory phase) was significantly enhanced after 2 days. Acid phosphatase secretion indicating macrophage activity (chronic inflammatory phase) was highest between 5 and 10 days after implantation. The secretion of both enzymes was low for the remaining periods of implantation of up to 6 months. Histologic examination revealed a discrete acute inflammatory phase, which progressed from acute to chronic between 10 and 15 days after implantation. The intensity of this chronic inflammatory reaction decreased with time from a mild reaction at 30 days to a more discrete response after 6 months. A thick collagenous capsule surrounding the implants had been formed by this time [53].

Blood Compatibility
When tested as substitutes of the infrarenal aorta in dogs, two out of four P3HO-3HH impregnated Dacron grafts were covered by a thick thrombotic matrix laying over the entire graft, with occasional islets of endothelial-like cells. The remaining two grafts were characterized by isolated small thrombi and the development of a thin collagenous capsule together with endothelial cell coverage [53]. Severe thrombus formation was also observed on all three leaflets made from porous P3HO-3HH films after 4 weeks of implantation in sheep [268]. In contrast, incomplete endothelialization, but no thrombus formation, was reported from another study testing porous P3HO-3HH heart valve scaffolds in sheep [280].

6.3
Biodegradability

In Vitro Degradation

Solution-cast films of P3HO-3HH (82% 3HO) were found to be very resistant to hydrolysis after 20 weeks of incubation in buffer solution at pH 10 and 37 °C. The slow decrease in molecular weight was associated with a strong increase in the crystallinity. Addition of PDLLA oligomer led to an acceleration of the P3HO-3HH degradation in these experiments [282]. No accelerating effect by blending with PLLA, PDLLA, or PEG was reported from a subsequent study. However, the introduction of polar carboxylic groups in side chains resulted in a significant increase in the degradation rate of P3HO-3HH due to enhanced water penetration into the polymer matrix [171]. The very slow hydrolysis of P3HO-3HH (97% 3HO) was confirmed in studies investigating the influence of enzymes on the P3HO-3HH degradation process. It was found that acid phosphatase and β-glucuronidase, both associated with the foreign-body reaction, are not involved in the polymer degradation [283]. Long-term in vitro studies on P3HO-3HH solution-cast films (97% HO) in either water (pH 7.0) or buffer solution (pH 7.4) at 37 °C revealed a simple hydrolytic degradation process characterized by low water absorption, slow gradual molecular weight loss, and negligible mass loss after 24 months of incubation [54]. At that time, the molecular weight of the films reached approximately 25% of the initial value in both water and buffer solution [53].

The polymer stability was also tested in terms of a suitable sterilization method. It was reported that the molecular weight M_w is decreased by 5% or 17% after sterilization by ethylene oxide or gamma-irradiation. Significant changes in both the physicochemical and tensile properties were observed after gamma-irradiation [284].

In Vivo Degradation

The degradation of P3HO-3HH in vivo was found to be slower in comparison to in vitro conditions in buffer solution, with less than 30% loss in the molecular weight 6 months after implantation s.c. in rats [53]. A molecular weight loss of less than 30% was also reported from in vivo experiments using porous P3HO-3HH heart valve scaffolds 17 weeks [280] or 24 weeks [268] after implantation in sheep. The degradation rate was found to be unaltered by increasing the surface area with porous P3HO-3HH scaffolds in comparison to dense films [268]. P3HO-3HH samples implanted s.c. in mice showed a 50% decrease in the molecular weight 40 weeks after implantation. No significant differences between the molecular weights at the surface and in the interior of the samples were observed in these experiments [3]. Based on the in vivo data it can be expected that the time for complete resorption of P3HO-3HH is in the order of magnitude of that for P3HB.

6.4
Applications in Tissue Engineering

Heart Valves

Initial studies have shown that biodegradable scaffolds made of polyglactin/ PGA mesh composites seeded with autologous vascular cells can be implanted to replace a single pulmonary valve leaflet in a lamb model [285]. However, attempts to construct a trileaflet heart valve scaffold failed due to stiffness of the materials [41]. In subsequent studies, P3HO-3HH was introduced as an elastomeric polymer to generate scaffolds with appropriate mechanical properties. These constructs first had a multilayered structure. The conduit wall consisted of a highly porous PGA mesh on the inside connected to a nonporous P3HO-3HH film on the outside. The valve leaflets had a PGA-(P3HO-3HH)-PGA sandwich structure with a P3HO-3HH nonporous film covered by a layer of PGA mesh on each side. The PGA mesh was chosen as a porous matrix for cell seeding and growth, while the P3HO-3HH layer provides flexibility and mechanical strength. These scaffolds were tested in vitro for cell attachment under pulsatile flow conditions [41]. However, the PGA-(P3HO-3HH)-PGA sandwich combination failed in vivo because of thrombus formation and lack of good tissue formation [268].

Modified constructs, consisting of the PGA-(P3HO-3HH)-PGA conduit and leaflets made from porous P3HO-3HH films were seeded with autologous vascular cells and implanted to replace the pulmonary valve and main pulmonary artery for up to 24 weeks in sheep. Postoperative examination of the preseeded scaffolds showed the formation of viable tissue without thrombus formation, while unseeded controls developed thrombi on all leaflets after 4 weeks. However, it was concluded that P3HO-3HH, due to the slow resorption, is not the ideal polymer for heart valve scaffolds [268]. The remodeling kinetics of these scaffolds after implantation have also been analyzed. While both the elastin and proteoglycans/glycosaminoglycans levels were comparable to those found in natural tissue within the 24-week period, there was a continuous increase in production and deposition of collagen. This might have been caused by a suboptimal matrix regulation, a pathologic collagen production, or both. Such pathologic matrix remodelling with excessive ECM deposition is not only a common problem after vascular injury but also one of the limiting factors for tissue engineering strategies to replace small-caliber vascular grafts [286].

Heart valve scaffolds with conduit and leaflets made entirely from P3HO-3HH porous films [287] were tested in vitro under static and pulsatile flow cell culture conditions [44, 281]. A larger number of vascular cells, together with the formation of an oriented and confluent cell layer, was found on scaffolds exposed to pulsatile flow [281]. The P3HO-3HH heart valves, preseeded with vascular cells under static conditions, were tested for up to 17 weeks in the pulmonary position in a lamb model. The valves showed good functionality with minimal regurgitation. All preseeded constructs were covered with tissue while the unseeded control showed no tissue formation and an inappropriate cell number and collagen content compared with native tissue. A limitation of the preseeded scaffolds is the lack of formation of a confluent endothelial cell layer, although no thrombi have been observed. Tensile tests showed a continuous decrease in both the scaffold modulus and strength during the implantation period. Based on the mechanical testing it was concluded that the scaffold is no longer necessary and should have been degraded at 17 weeks after implantation [280]. More recently, the application of rapid prototyping techniques to fabricate porous P3HO-3HH heart valve scaffolds has been described [269].

Vascular Grafts

Tubular scaffolds made from a polyglactin/PGA mesh composite were successfully tested to replace a pulmonary artery in sheep [288]. However, preliminary studies using this material in the high pressure systemic circulation resulted in aneurysm formation due to the fast polymer resorption within 6–8 weeks. Therefore, P3HO-3HH was introduced as an alternative material with a much longer degradation time to withstand the systemic pressure [40]. Vascular grafts made from a PGA mesh as inner layer, to promote cell attachment and tissue formation, and three outer layers of nonporous P3HO-3HH films, to provide mechanical support, were tested in the aortic position in sheep. Scaffolds preseeded with a mixture of endothelial cells, smooth muscle cells, and fibroblasts remained without aneurysm formation during the 5-month implantation period. In contrast, all unseeded control grafts became occluded, which was in part caused by the thrombogenic PGA mesh. A very thin and loose tissue was found on the unseeded materials. Increased cell density and collagen formation as well as endothelial cell layering has been found on the preseeded materials. However, metabolic activities tended to increase significantly over time [40]. As described above, this might result in excessive ECM production [286], which has to be clarified in a long-term implantation study.

It is worth noting that P3HO-3HH has also been tested as a sealant of Dacron vascular prostheses. Although the polymer impregnation contributed to the impermeabilization of the graft wall the P3HO-3HH impregnated grafts showed a delayed healing process in comparison to preclotted nonimpregnated grafts when tested as an infrarenal aortic substitute in dogs. This was explained by the slow resorption time of P3HO-3HH limiting collagenous tissue formation and infiltration into the graft [53].

7
Outlook

The development of biomaterials is one of the most promising fields in biomedical research. Although an increasing number of biodegradable polymers have been examined over the past 40 years, materials with optimal properties which allow for an appropriate host response to ensure functional repair of damaged or diseased tissue and organs are still not available for many clinical applications. Thus, even well-established synthetically produced biodegradable polyesters such as PGA, PDLLA, PLLA, and their copolymers, which are approved for a variety of medical applications, are not unrivaled with respect to alternative biomaterials. Ultimately, the material with the most appropriate properties will become accepted in a specific biomedical application. The material's processability, mechanical, degradation, and especially biocompatibility properties, as well as other attributes relevant to medical implant design (such as sterilizability and sutureability) will need to be evaluated in context of the application.

Potential uses of P3HB in tissue engineering could be as implants that require a longer retention time or a higher stability towards the surrounding environment, but which eventually absorb. Examples of applications could be in the repair of bone, nerve, the cardiovascular system, or urinary and gastrointestinal tract defects. P4HB is promising for tissue engineering applications in the heart and circulatory system. Additionally, the polymer may replace presently used materials having shorter degradation times, such as suture materials or surgical textiles. Elastomeric PHAs, such as P3HO-3HH or P3HB-4HB, could be of interest in the traditional biomedical applications of elastomers [289]. They also have particularly favorable mechanical properties when used as scaffolds for repair and regeneration of elastomeric tissue, such as in the cardiovascular system. Device Master Files for P4HB and P3HB-4HB have recently been submitted by Tepha Inc. to the FDA, facilitating the development of medical products based on these polymers. FDA clearance of the first PHA device, a P4HB based suture, can be expected by 2006 (Martin DP, personal communication).

References

1. Vacanti JP, Langer R (1999) Lancet 354:I32
2. Leor J, Cohen S (2004) Ann NY Acad Sci 1015:312
3. Williams SF, Martin DP, Horowitz DM, Peoples OP (1999) Int J Biol Macromol 25:111
4. Vainionpää S, Rokkanen P, Törmälä P (1989) Prog Polym Sci 49:679
5. Hollinger JO (ed) (1995) Biomedical applications of synthetic biodegradable polymers. CRC, Boca Raton
6. Amass W, Amass A, Tighe B (1998) Polym Int 47:89
7. Ueda H, Tabata Y (2003) Adv Drug Deliv Rev 55:501
8. Liu DM, Dixit V (eds) (1997) Porous materials for tissue engineering. Trans Tech, Uetikon-Zürich
9. Atala A, Mooney D (eds) (1997) Synthetic biodegradable polymer scaffolds. Birkhaeuser, Boston
10. Seal BL, Otero TC, Panitch A (2001) Mater Sci Eng R34:147
11. Atala A, Lanza RP (eds) (2002) Methods of tissue engineering. Academic, San Diego

12. Hocking PJ, Marchessault RH (1994) In: Griffin GJL (ed) Chemistry and technology of biodegradable polymers. Blackie, London, p 48
13. Hasirci V (2000) In: Wise DL (ed) Biomaterials and bioengineering handbook. Marcel Dekker, New York, p 141
14. Zinn M, Witholt B, Egli T (2001) Adv Drug Deliv Rev 53:5
15. Asrar J, Hill JC (2002) J Appl Polym Sci 83:457
16. Williams SF, Martin DP (2002) In: Doi Y, Steinbüchel A (eds) Biopolymers, vol 4. Polyesters III, applications and commercial products. Wiley, Weinheim, p 91
17. Freier T, Sternberg K, Behrend D, Schmitz KP (2003) In: Steinbüchel A (ed) Biopolymers, vol 10. General aspects and special applications. Wiley, Weinheim, p 247
18. Dawes EA, Senior PJ (1973) In: Rose AH, Tempest DW (eds) Advances in microbial physiology, vol 10. Academic, London, p 203
19. Doi Y (1990) Microbial polyesters. Wiley, Weinheim
20. Lee SY (1996) Biotechnol Bioeng 49:1
21. Madison LL, Huisman GW (1999) Microbiol Mol Biol Rev 63:21
22. Sudesh K, Abe H, Doi Y (2000) Prog Polym Sci 25:1503
23. Steinbüchel A (2001) Macromol Biosci 1:1
24. Steinbüchel A, Valentin HE (1995) FEMS Microbiol Lett 128:219
25. Baptist JN (1962) US 3044942
26. Baptist JN, Ziegler JB (1965) US 3225766
27. Korsatko W, Wabnegg B, Braunegg G, Lafferty RM, Strempfl F (1983) Pharm Ind 45:525
28. Korsatko W, Wabnegg, B, Tillian HM, Braunegg G, Lafferty RM (1983) Pharm Ind 45:1004
29. Korsatko W, Wabnegg B, Tillian HM, Egger G, Pfragner R, Walser V (1984) Pharm Ind 46:952
30. Bonfield W (1988) J Biomed Eng 10:522
31. Heimerl A, Pietsch H, Rademacher KH, Schwengler H, Winkeltau G, Treutner KH (1989) EP 0336148
32. Bowald SF, Johansson EG (1990) EP 0349505
33. Duvernoy O, Malm T, Ramström J, Bowald S (1995) Thorac Cardiovasc Surg 43:271
34. Hazari A, Johansson-Ruden G, Junemo-Bostrom K, Ljungberg C, Terenghi G, Green C, Wiberg M (1999a) J Hand Surg 24:291
35. Hazari A, Wiberg M, Johanssson-Ruden G, Green C, Terenghi G (1999) Brit J Plast Surg 52:653
36. Ljungberg C, Johansson-Ruden G, Bostrom KJ, Novikov L, Wiberg M (1999) Microsurg 19:259
37. Novikov LN, Novikova LN, Mosahebi A, Wiberg M, Terenghi G, Kellerth JO (2002) Biomaterials 23:3369
38. Freier T, Kunze C, Nischan C, Kramer S, Sternberg K, Saß M, Hopt UT, Schmitz KP (2002) Biomaterials 23:2649
39. Kunze C, Bernd HE, Androsch R, Nischan C, Freier T, Kramer S, Kramp B, Schmitz KP (2006) Biomaterials 27:192
40. Shum-Tim D, Stock U, Hrkach J, Shinoka T, Lien J, Moses MA, Stamp A, Taylor G, Moran AM, Landis W, Langer R, Vacanti JP, Mayer Jr JE (1999) Ann Thorac Surg 68:2298
41. Sodian R, Sperling JS, Martin DP, Stock U, Mayer JE Jr, Vacanti JP (1999) Tissue Eng 5:489
42. Stock UA, Sakamoto T, Hatsuoka S, Martin DP, Nagashima M, Moran AM, Moses MA, Khalil PN, Schoen FJ, Vacanti JP, Mayer JE (2000) J Thorac Cardiovasc Surg 120:1158

43. Hoerstrup SP, Sodian R, Daebritz S, Wang J, Bacha EA, Martin DP, Moran AM, Guleserian KJ, Sperling JS, Kaushal S, Vacanti JP, Schoen FJ, Mayer JE (2000) Circulation 102:III44
44. Sodian R, Hoerstrup SP, Sperling JS, Martin DP, Daebritz S (2000) ASAIO J 46:107
45. Hoerstrup SP, Zünd G, Sodian R, Schnell AM, Grünenfelder J, Turina MI (2001) Eur J Cardiothorac Surg 20:164
46. Opitz F, Schenke-Layland K, Richter W, Martin DP, Degenkolbe I, Wahlers T, Stock UA (2004) Ann Biomed Eng 32:212
47. Martin DP, Williams SF (2003) Biochem Eng J 16:97
48. Deng Y, Zhao K, Zhang XF, Hu P, Chen GQ (2002) Biomaterials 23:4049
49. Wang YW, Wu Q, Chen GQ (2004) Biomaterials 25:669
50. Chung CW, Kim HW, Kim YB, Rhee YH (2003) Int J Biol Macromol 32:17
51. Freier T, Backhaus-Pohl C, Steinhoff G, Schmitz KP (2002) Ann Meet German Soc Biomat. Dresden, Germany
52. Dufresne A, Vincendon M (2000) Macromolecules 33:2998
53. Marois Y, Zhang Z, Vert M, Deng X, Lenz RW, Guidoin R (2000) In: Mauli Agrawal C, Parr JE, Lin ST (eds) Synthetic bioabsorbable polymers for implants. American Society for Testing and Materials, West Conshohocken, p 12
54. Marois Y, Zhang Z, Vert M, Deng X, Lenz R, Guidoin R (2000) J Biomed Mater Res 49:216
55. Gogolewski S, Jovanovic M, Perren SM, Dillon JG, Hughes MK (1993) J Biomed Mater Res 27:1135
56. Martin DP, Skraly FA, Williams SF (1999) WO 99/32536
57. Satkowski MM, Melik DH, Autran JP, Green PR, Noda I, Schechtman LA (2002) In: Doi Y, Steinbüchel A (eds) Biopolymers, vol 3b. Polyesters II, properties and chemical synthesis. Wiley, Weinheim, p 231
58. Gagnon KD, Fuller RC, Lenz RW, Farris RJ (1992) Rubber World 207:32
59. Webb AR, Yang J, Ameer GA (2004) Exp Op Biol Ther 4:801
60. Holmes PA (1988) In: Bassett DC (ed) Developments in crystalline polymers. Elsevier, London, p 1
61. Engelberg I, Kohn J (1991) Biomaterials 12:292
62. Doi Y, Segawa A, Kunioka M (1990) Int J Biol Macromol 12:106
63. Doi Y, Kitamura S, Abe H (1995) Macromolecules 28:4822
64. Avella M, Martuscelli E, Raimo M (2000) J Mater Sci 35:523
65. Scandola M, Ceccorulli G, Pizzoli M, Gazzano M (1992) Macromolecules 25:1405
66. Luo S, Grubb DT, Netravali AN (2002) Polymer 43:4159
67. Renstadt R, Karlsson S, Albertsson AC (1997) Polym Degr Stab 57:331
68. Ishikawa K, Kawaguchi Y, Doi Y (1991) Kobunshi Ronbunshu 48:221
69. Savenkova L, Gercberga Z, Nikolaeva V, Dzene A, Bibers I, Kalnin M (2000) Proc Biochem 35:573
70. Baltieri RC, Innicentini Mei LH, Bartoli J (2003) Macromol Symp 197:33
71. Kunze C, Freier T, Kramer S, Schmitz KP (2002) J Mater Sci Mater Med 13:1051
72. Freier T, Schmitz KP (2002) Int Symp Biol Polyesters. Münster, Germany
73. Hull EH (1990) In: Medical plastics today & tomorrow. Society of the Plastics Industry, Anaheim, p 1
74. Labreque LV, Kumar RA, Davé V, Gross RA, McCarthy SP (1997) J Appl Polym Sci 66:1507
75. Hammond T, Liggat JJ, Montador JH, Webb A (1994) WO 94/28061
76. Ghiya VP, Davé V, Gross RA, McCarthy SP (1995) Polym Prepr 36:420
77. Asrar J, Pierre JR (2000) US 6127512

78. Choi JS, Park WH (2004) Polym Test 23:455
79. Freier T, Kunze C, Schmitz KP (2001) J Mater Sci Lett 20:1929
80. Kumagai Y, Doi Y (1992) Makromol Chem Rapid Commun 13:179
81. Abe H, Matsubara I, Doi Y, Hori Y, Yamaguchi A (1994) Macromolecules 27:6018
82. Abe H, Matsubara I, Doi Y (1995) Macromolecules 28:844
83. Scandola M, Focarete ML, Adamus G, Sikorska W, Baranowska I, Swierczek S, Gnatowski M, Kowalczuk M, Jedlinski Z (1997) Macromolecules 30:2568
84. El-Taweel SH, Stoll B, Höhne GWH, Mansour AA, Seliger H (2004) J Appl Polym Sci 94:2528
85. Kumagai Y, Doi Y (1992b) Polym Deg Stab 36:241
86. Gassner F, Owen AJ (1994) Polymer 35:2233
87. Immirzi B, Malinconico M, Orsello G, Portofino S, Volpe MG (1999) J Mater Sci 34:1625
88. Abe H, Doi Y, Kumagai Y (1994) Macromolecules 27:6012
89. Zhao K, Deng Y, Chen GQ (2003) Biochem Eng J 16:115
90. Zhao K, Deng Y, Chen JC, Chen GQ (2003) Biomaterials 24:1041
91. Kusaka S, Iwata T, Doi Y (1998) J Macromol Sci A35:319
92. Park JW, Doi Y, Iwata T (2004) Biomacromolecules 5:1557
93. Doyle C, Tanner ET, Bonfield W (1991) Biomaterials 12:841
94. Wang YW, Yang F, Wu Q, Cheng YC, Yu PHF, Chen J, Chen GQ (2005) Biomaterials 26:755
95. Wang M, Wenig J, Ni J, Goh CH, Wang CX (2001) Key Eng Mater 192–195:741
96. Williams DF (ed) (1999) The Williams dictionary of biomaterials. Liverpool University Press, Liverpool, p 40
97. Reusch RN, Sparrow AW, Gardiner J (1992) Biochim Biophys Acta 1123:33
98. Saito T, Tomita K, Juni K, Ooba K (1991) Biomaterials 12:309
99. Taylor MS, Daniels AU, Andriano KP, Heller J (1994) J Appl Biomater 5:151
100. Chaput C, Yahia L, Selmani A, Rivard CH (1995) In: Mikos AG, Leong KW, Yaszemski MJ, Tamada JA, Radomsky ML (eds) Polymers in medicine and pharmacy. Materials Research Society, Pittsburgh, p 111
101. Chaput C, Assad M, Yahia LH, Rivard CH, Selmani A (1995) Biomater Living Syst Interact 3:29
102. Shishatskaya EI, Volova TG (2004) J Mater Sci Mater Med 15:915
103. Cheng G, Cai Z, Wang L (2003) J Mater Sci Mater Med 14:1073
104. Schmack G, Jehnichen D, Vogel R, Tandler B (2000) J Polym Sci B38:2841
105. Schmack G, Kramer S, Dorschner H, Gliesche K (2004) Polym Deg Stab 83:467
106. Rivard CH, Chaput CJ, DesRosiers EA, Yahia LH, Selmani A (1995) J Appl Biomater 6:65
107. Chaput C, Yahia LH, Landry D, Rivard CH, Selmani A (1995) Biomater Living Syst Interact 3:19
108. Giavaresi G, Tschon M, Daly JH, Liggat JJ, Fini M, Torricelli P, Giardino R (2004) Int J Artif Org 27:796
109. Dang MH, Birchler F, Wintermantel E (1997) Polym Degr Stab 5:49
110. Zhang DM, Cui FZ, Luo ZS, Lin YB, Zhao K, Chen GQ (2000) Surf Coat Tech 131:350
111. Hasirci V, Tezcaner A, Hasirci N, Süzer S (2003) J Appl Polym Sci 87:1285
112. Lee SJ, Lee YM, Khang G, Kim IY, Lee B, Lee HB (2002) Macromol Res 10:150
113. Etzrodt D, Rybka M, Röpke C, Michalik I, Behrend D, Schmitz KP (1997) Biomed Tech 42(suppl 1):445
114. Nitschke M, Schmack G, Janke A, Simon F, Pleul D, Werner C (2002) J Biomed Mater Res 59:632
115. Yang X, Zhao K, Chen GQ (2002) Biomaterials 23:1391

116. Zhao K, Yang X, Chen GQ, Chen JC (2002) J Mater Sci Mater Med 13:849
117. Wang YW, Wu Q, Chen GQ (2003) Biomaterials 24:4621
118. Hu SG, Jou CH, Yang MC (2003) Biomaterials 24:2685
119. Pouton CW, Kennedy JE, Notarianni LJ, Gould PL (1988) In: Proc Int Symp Contr Rel Bioact Mater. The Controlled Release Society, Lincolnshire, p 179
120. Piddubnyak V, Kurcok P, Matuszowicz A, Glowala M, Fiszer-Kierzkowska A, Jedlinski Z, Juzwa M, Krawczyk Z (2004) Biomaterials 25:5271
121. Saad B, Ciardelli G, Matter S, Welti M, Uhlschmid GK, Neuenschwander P, Suter UW (1996) J Mater Sci Mater Med 7:56
122. Rouxhet L, Legras R, Schneider YJ (1998) Macromol Symp 130:347
123. Nebe B, Forster C, Pommerenke H, Fulda G, Behrend D, Berneweski U, Schmitz KP, Rychly J (2001) Biomaterials 22:2425
124. Wang YW, Wu Q, Chen GQ (2005) Biomaterials 26:899
125. Gurav N, Downes S (1994) J Mater Sci Mater Med 5:784
126. Torun Köse G, Ber S, Korkusuz F, Hasirci V (2003) J Mater Sci Mater Med 14:121
127. Torun Köse G, Kenar H, Hasirci N, Hasirci V (2003) Biomaterials 24:1949
128. Tesema Y, Raghavan D, Stubbs J III (2004) J Appl Polym Sci 93:2445
129. Zheng Z, Bei FF, Tian HL, Chen GQ (2005) Biomaterials 26:3537
130. Deng Y, Lin XS, Zheng Z, Deng JG, Chen JC, Ma H, Chen GQ (2003) Biomaterials 24:4273
131. Zheng Z, Deng Y, Lin XS, Zhang LX, Chen GQ (2003) J Biomater Sci Polym Edn 14:615
132. Rivard CH, Chaput C, Rhalmi S, Selmani A (1996) Ann Chir 50:651
133. Lee IS, Kwon OH, Meng W, Kang IK, Ito Y (2004) Macromol Res 12:374
134. Davies S, Tighe B (1995) Polym Prepr 36:103
135. Foster LJR, Davies SM, Tighe BJ (2001) J Biomater Sci Polym Ed 12:317
136. Ostwald J, Dommerich S, Nischan C, Kramp B (2003) Laryngo-Rhino-Otol 82:693
137. Tezcaner A, Bugra K, Hasirci V (2003) Biomaterials 24:4573
138. Schué F, Clarotti G, Sledz J, Mas A, Geckeler KE, Göpel W, Orsetti A (1993) Makromol Chem Macromol Symp 73:217
139. Clarotti G, Sledz F, Schué J, Ait Ben Aoumar A, Geckeler KE, Orsetti A, Paleirac G (1992) Biomaterials 13:832
140. Najimi A, Crespy S, Mas A, Schue F (1999) Innov Techn Biol Med 20:187
141. Teese M, Pompe T, Nitschke M, Herold N, Zimmermann R, Werner C (2004) World Biomat Congr. Sydney, Australia
142. Malm T, Bowald S, Bylock A, Busch C (1992c) J Thorac Cardiovasc Surg 104:600
143. Malm T, Bowald S (1994) J Thorac Cardiovasc Surg 107:628
144. Malm T, Bowald S, Karacagil S, Bylock A, Busch C (1992a) Scand J Thorac Cardiovasc Surg 26:9
145. Malm T, Bowald S, Bylock A, Busch C, Saldeen T (1994) Eur Surg Res 26:298
146. Giavaresi G, Tschon M, Borsari V, Daly JH, Liggat JJ, Fini M, Bonazzi V, Nicolini A, Carpi A, Morra M, Cassinelli C, Giardino R (2004) Biomed Pharmacother 58:411
147. Volova T, Shishatskaya E, Sevastianov V, Efremov S, Mogilnaya O (2003) Biochem Eng J 16:125
148. Shishatskaya EI, Volova TG, Puzyr AP, Mogilnaya OA (2004) J Mater Sci Mater Med 15:719
149. Saß M (1996) MD thesis, University of Rostock
150. Löbler M, Saß M, Kunze C, Schmitz KP, Hopt UT (2002) Biomaterials 23:577
151. Löbler M, Saß M, Kunze C, Schmitz KP, Hopt UT (2002) J Biomed Mater Res 61:165
152. Knowles JC, Hastings GW (1993) J Mater Sci Mater Med 4:102

153. Herold A, Bruch HP, Weckbach A, Romen W, Schönefeld G (1988) In: Pannike A (ed) Hefte zur Unfallheilkunde 200. Springer, Berlin Heidelberg New York, p 665
154. Sadowski B (1988) MD thesis, University of Würzburg
155. Luklinska ZB, Bonfield W (1997) J Mater Sci Mater Med 8:379
156. Leenstra TS, Maltha JC, Kuijpers-Jagtman AM (1995) J Mater Sci Mater Med 6:445
157. Gotfredsen K, Nimb L, Hjorting-Hansen E (1994) Clin Oral Impl Res 5:83
158. van der Giessen W, Lincoff AM, Schwartz RS, van Beusekom HMM, Serruys PW Jr, Holmes DR, Ellis SG, Topol EJ (1996) Circulation 94:1690
159. Unverdorben M, Spielberger A, Schywalski M, Labahn D, Hartwig S, Schneider M, Lootz D, Behrend D, Schmitz K, Degenhardt R, Schaldach M, Vallbracht C (2002) Intervent Radiol 25:127
160. Lootz D, Behrend D, Kramer S, Freier T, Haubold A, Benkießer G, Schmitz KP, Becher B (2001) Biomaterials 22:2447
161. Iordanskii AL, Dmitriev EV, Kamaev PP, Zaikov GE (1999) J Appl Polym Sci 74:595
162. Coussot-Rico P, Clarotti G, Ait Ben Aoumar A, Najimi A, Sledz J, Schué F, Quatrefages R (1994) Eur Polym J 30:1327
163. Zinner G, Behrend D, Schmitz KP (1998) Biomed Tech 43(suppl 1):432
164. Zinner G, Behrend D, Schmitz KP (1997) Biomed Tech 42(suppl 1):55
165. Sevastianov VI, Perova NV, Shishatskaya EI, Kalacheva GS, Volova TG (2003) J Biomater Sci Polym Edn 14:1029
166. Scholz C (2000) In: Scholz C (ed) Polymers from renewable resources: biopolyesters and biocatalysis. American Chemical Society, Washington, p 328
167. Doi Y, Kanesawa Y, Kawaguchi Y, Kunioka M (1989) Makromol Chem Rapid Commun 10:227
168. Zhang L, Xiong C, Deng X (1995) J Appl Polym Sci 56:103
169. Kurcok P, Kowalczuk M, Adamus G, Jedlinski Z, Lenz RW (1995) J Macromol Sci A32:875
170. Abe H, Doi Y, Satkowski MM, Noda I (1994) In: Doi Y, Fukuda K (eds) Biodegradable plastics and polymers. Elsevier, Amsterdam, p 591
171. Renard E, Walls M, Guerin P, Langlois V (2004) Polym Deg Stab 85:779
172. Atkins TW, Peacock SJ (1997) J Microencapsul 14:35
173. Tokiwa Y, Suzuki T, Takeda K (1986) Agric Biol Chem 50:1323
174. Jendrossek D, Schirmer A, Schlegel HG (1996) Appl Microbiol Biotechnol 46:451
175. Winkler FK, D'Arcy A, Hunziker W (1990) Nature 343:771
176. Yasin M, Tighe BJ (1993) Plast Rubber Compos Proc Appl 19:15
177. Holland SJ, Jolly AM, Yasin M, Tighe BJ (1987) Biomaterials 8:289
178. Yasin M, Holland SJ, Tighe BJ (1990) Biomaterials 11:451
179. Yasin M, Holland SJ, Jolly AM, Tighe BJ (1989) Biomaterials 10:400
180. Yasin M, Tighe BJ (1992) Biomaterials 13:9
181. Kishida A, Yoshioka S, Takeda Y, Uchiyama M (1989) Chem Pharm Bull 37:1954
182. Yoshioka S, Kishida A, Izumikawa S, Aso Y, Takeda Y (1991) J Control Release 16:341
183. Knowles JC, Hastings GW (1992) J Mater Sci Mater Med 3:352
184. Jones NL, Cooper JJ, Waters RD, Williams DF (2000) In: Agrawal CM, Parr JE, Lin ST (eds) Synthetic bioabsorbable polymers for implants. American Society for Testing and Materials, West Conshohocken, p 69
185. Eldsäter C, Albertsson AC, Karlsson S (1997) Acta Polym 48:478
186. Renstadt R, Karlsson S, Albertsson AC (1999) Polym Degrad Stab 63:201
187. Kanesawa Y, Doi Y (1990) Makromol Chem Rapid Commun 11:679
188. Foster LJR, Tighe BJ (2005) Polym Deg Stab 87:1
189. Li H, Chang J (2005) Polym Deg Stab 87:301

190. Lootz D, Kobow M, Zinner G, Michalik I, Behrend D, Schmitz KP (1997) Biomed Tech 42(Suppl 1):53
191. Mitomo H, Watanabe Y, Ishigaki I, Saito T (1994) Polym Degr Stab 45:11
192. Ishikawa K (1996) US 5480394
193. Carswell-Pomerantz T, Dong L, Hill DJT, O'Donnell JH, Pomery PJ (1996) In: Clough RL, Shalaby SW (eds) Irradiation of polymers: fundamentals and technological applications. American Chemical Society, Washington, p 11
194. Luo S, Netravali AN (1999) J Appl Polym Sci 73:1059
195. Bibers I, Kalnins M (1999) Mech Compos Mater 35:169
196. Miyazaki SS, Yep AR, Kolton F, Hermida EB, Povolo F, Fernandez EG, Chiellini E (2003) Macromol Symp 197:57
197. Bledzki AK, Gassan J, Heyne M (1994) Angew Makromol Chem 219:11
198. Koosha F, Muller RH, Davis SS (1989) CRC Crit Rev Therap Drug Carrier Syst 6:117
199. Frazza EJ, Schmitt EE (1971) J Biomed Mater Res Biomed Mater Symp 1:43
200. Kunioka M, Tamaki A, Doi Y (1989) Macromolecules 22:694
201. Behrend D, Schmitz KP, Haubold A (2000) Adv Eng Mater 2:123
202. Miller ND, Williams DF (1987) Biomaterials 8:129
203. Scherer MA, Früh HJ, Ascherl R, Mau H, Siebels W, Blümel G (1992) In: Planck H, Dauner M, Renardy M (eds) Degradation phenomena on polymeric biomaterials. Springer, Berlin Heidelberg New York, p 77
204. Kalangos A, Faidutti B (1996) J Thorac Cardiovasc Surg 112:1401
205. Behrend D, Lootz D, Schmitz KP, Schywalski M, Labahn D, Hartwig S, Schaldach M, Unverdorben M, Vallbracht C, Laenger F (1998) Am J Cardiol 82(special issue):4S
206. Unverdorben M, Schywalski M, Labahn D, Hartwig S, Laenger F, Lootz D, Behrend D, Schmitz KP, Schaldach M, Vallbracht C (1998) Am J Cardiol 82(special issue):5S
207. Olsen L, Bowald S, Busch C, Carlsten J, Eriksson I (1992) Scand J Urol Nephrol 26:323
208. Kramp B, Bernd HE, Schumacher WA, Blynow M, Schmidt W, Kunze C, Behrend D, Schmitz KP (2002) Laryngo-Rhino-Otol 81:351
209. Vainionpää S, Vihtonen K, Mero M, Pätiälä H, Rokkanen P, Kilpikari J, Törmälä P (1986) Acta Orthopaed Scand 57:237
210. Kostopoulos L, Karring T (1994) Clin Oral Impl Res 5:66
211. Malm T, Bowald S, Bylock A, Saldeen T, Busch C (1992b) Scand J Thorac Cardiovasc Surg 26:15
212. Nkere UU, Whawell SA, Sarraf CE, Schofield JB, O'Keefe PA (1998) Thorac Cardiovasc Surg 46:77
213. Lindström L, Wigström L, Dahlin LG, Aren C, Wranne B (2000) Scand Cardiovasc J 34:331
214. Schmack G, Gliesche K, Nitschke M, Werner C (2002) Biomaterialien 3:21
215. Winkeltau GJ, Treutner KH, Kleimann E, Lerch MM, Ger R, Haase G, Schumpelick V (1993) Dis Colon Rect 36:154
216. Mosahebi A, Woodward B, Wiberg M, Martin R, Terenghi G (2001) Glia 34:8
217. Mosahebi A, Fuller P, Wiberg M, Terenghi G (2002) Exp Neurol 173:213
218. Mosahebi A, Wiberg M, Terenghi G (2003) Tissue Eng 9:209
219. McKay Hart A, Wiberg M, Terenghi G (2003) Br J Plast Surg 56:444
220. Young RC, Wiberg M, Terenghi G (2002) Br J Plast Surg 55:235
221. Mohanna PN, Young RC, Wiberg M, Terenghi G (2003) J Anat 203:553
222. Knowles JC, Hastings GW (1991) Biomaterials 12:210
223. Vasenius J, Vainionpää S, Vihtonen K, Mero M, Mikkola J, Pellinen M, Rokkanen P, Törmälä P (1991) In: Barbosa MA (ed) Biomaterials degradation. North-Holland, Amsterdam, p 393

224. Fukada E, Ando Y (1988) Biorheology 25:297
225. Knowles JC, Mahmud FA, Hastings GW (1991) Clin Mater 8:155
226. Holmes PA (1985) Phys Technol 16:32
227. Kostopoulos L, Karring T (1994b) Clin Oral Impl Res 5:75
228. Galego N, Rozsa C, Sanchez R, Fung J, Vazquez A, Tomas JS (2000) Polym Test 19:485
229. Boeree NR, Dove J, Cooper JJ, Knowles J, Hastings GW (1993) Biomaterials 14:793
230. Ni J, Wang M (2002) Mater Sci Eng C20:101
231. Chen LJ, Wang M (2002) Biomaterials 23:2631
232. Li H, Chang J (2004) Biomaterials 25:5473
233. Saad B, Kuboki Y, Welti M, Uhlschmid GK, Neuenschwander P, Suter UW (2000) Artif Org 24:939
234. Webb A, Adsetts JR (1986) GB 2166354
235. Schaffer J, Voigt HD, Rauchstein KD, Wengemuth K (1995) DE 4416357
236. Foster LJR, Tighe BJ (1994) J Environ Polym Degr 2:185
237. Nakamura S, Doi Y, Scandola M (1992) Macromolecules 25:4237
238. Steinbüchel A, Valentin HE, Schönebaum A (1994) J Environ Polym Degr 2:67
239. Kimura H, Ohura T, Takeishi M, Nakamura S, Doi Y (1999) Polym Int 48:1073
240. Sudesh K, Fukui T, Taguchi K, Iwata T, Doi Y (1999) Int J Biol Macromol 25:79
241. Choi MH, Yoon SC, Lenz RW (1999) Appl Environm Microbiol 65:1570
242. Hein S, Söhling B, Gottschalk G, Steinbüchel A (1997) FEMS Microbiol Lett 153:411
243. Song S, Hein S, Steinbüchel A (1999) Biotechn Lett 21:193
244. Kunioka A, Nakamura Y, Doi Y (1988) Polym Commun 29:174
245. Kimura H, Yoshida Y, Doi Y (1992) Biotechn Lett 14:445
246. Saito Y, Doi Y (1994) Int J Biol Macromol 16:99
247. Grabow N, Schmohl K, Khosravi A, Philipp M, Scharfschwerdt M, Graf B, Stamm C, Haubold A, Schmitz KP, Steinhoff G (2004) Artif Org 28:971
248. Nelson T, Kaufman E, Kline J, Sokoloff L (1981) J Neurochem 37:1345
249. Fuller DE, Hornfeldt CS (2003) Pharmacotherapy 23:1205
250. Ropero-Miller JD, Goldberger BA (1998) Clin Lab Med 18:727
251. Opitz F, Melle C, Schenke-Layland K, Degenkolbe I, Martin DP, von Eggeling F, Wahlers T, Stock UA (2004) Tissue Eng 10:611
252. Rabkin E, Hoerstrup SP, Aikawa M, Mayer JE, Schoen FJ (2002) J Heart Valve Dis 11:308
253. Nasseri BA, Pomerantseva I, Kaazempur-Mofrad MR, Sutherland FWH, Perry T, Ochoa E, Thompson CA, Mayer JE, Oesterle SN, Vacanti JP (2003) Tissue Eng 9:291
254. Dvorin EL, Wylie-Sears J, Kaushal S, Martin DP, Bischoff J (2003) Tissue Eng 9:487
255. Hoerstrup SP, Kadner A, Melnitchouk S, Trojan A, Eid K, Tracy J, Sodian R, Visjager JF, Kolb SA, Grunenfelder J, Zund G, Turina MI (2002) Circulation 106:I143
256. Kadner A, Hoerstrup SP, Zund G, Eid K, Maurus C, Melnitchouk S, Grunenfelder J, Turina MI (2002) Eur J Cardio-Thorac Surg 21:1055
257. Perry T, Kaushal S, Sutherland FWH, Guleserian KJ, Bischoff J, Sacks M, Mayer JE (2003) Ann Thorac Surg 75:761
258. Fu P, Sodian R, Lüders C, Lemke T, Kraemer L, Hübler M, Weng Y, Hoerstrup SP, Meyer R, Hetzer R (2004) ASAIO J 50:9
259. Hoerstrup SP, Kadner A, Breymann C, Maurus CF, Guenter CI, Sodian R, Visjager JF, Zund G, Turina MI (2002) Ann Thorac Surg 74:46
260. Kaviani A, Perry TE, Barnes CM, Oh JT, Ziegler MM, Fishman SJ, Fauza DO (2002) J Pediatr Surg 37:995
261. Fuchs JR, Pomerantseva I, Ochoa ER, Vacanti JP, Fauza DO (2003) J Pediatr Surg 38:1348

262. Opitz F, Schenke-Layland K, Cohnert TU, Starcher B, Halbhuber KJ, Martin DP, Stock UA (2004) Cardiovasc Res 63:719
263. Treckmann N (2005) MD thesis, University of Rostock
264. Korkusuz F, Korkusuz P, Eksioglu F, Gürsel I, Hasirci V (2001) J Biomed Mater Res 55:217
265. Drechsel A (2005) MD thesis, University of Rostock
266. Franconi F, Miceli M, Alberti L, Boatto G, Coinu R, de Montis MG, Tagliamonte A (2001) Thromb Res 102:255
267. Doi Y, Kanesawa Y, Kunioka M, Saito T (1990) Macromolecules 23:26
268. Stock UA, Nagashima M, Khalil PN, Nollert GD, Herden T, Sperling JS, Moran A, Lien J, Martin DP, Schoen FJ, Vacanti JP, Mayer JE Jr (2000) J Thorac Cardiovasc Surg 119:732
269. Sodian R, Loebe A, Martin DP, Hoerstrup SP, Potapov EV, Hausmann H, Lueth T, Hetzer R (2002) ASAIO J 48:12
270. Engelmayr GC Jr, Hildebrand DK, Sutherland FWH, Mayer JE Jr, Sacks MS (2003) Biomaterials 24:2523
271. Khosravi A, Stamm C, Philipp M, Freier T, Schmitz KP, Haubold A, Steinhoff G (2003) Mater Sci Forum 426:3067
272. Stamm C, Khosravi A, Grabow N, Schmohl K, Treckmann N, Drechsel A, Nan M, Schmitz KP, Haubold A, Steinhoff G (2004) Ann Thorac Surg 78:2084
273. Lee SH, Oh DH, Ahn WS, Lee Y, Choi J, Lee SY (2000) Biotechnol Bioeng 67:240
274. Asrar J, Valentin HE, Berger PA, Tran M, Padgette SR, Garbow JR (2002) Biomacromolecules 3:1006
275. Chen GQ, Zhang G, Park SJ, Lee SY (2001) Appl Microbiol Biotechnol 57:50
276. Fischer JJ, Aoyagi Y, Enoki M, Doi Y, Iwata T (2004) Polym Deg Stab 83:453
277. Hassan MK, Abdel-Latif SA, El-Roudi OM, Sharaf MA, Noda I, Mark JE (2004) J Appl Polym Sci 94:2257
278. Noda I, Satkowski MM, Dowrey AE, Marcott C (2004) Macromol Biosci 4:269
279. Yang M, Zhu S, Chen Y, Chang Z, Chen G, Gong Y, Zhao N, Zhang X (2004) Biomaterials 25:1365
280. Sodian R, Hoerstrup SP, Sperling JS, Daebritz S, Martin DP, Moran AM, Kim BS, Schoen FJ, Vacanti JP, Mayer JE Jr (2000) Circulation 102:III22
281. Sodian R, Hoerstrup SP, Sperling JS, Daebritz SH, Martin DP, Schoen FJ, Vacanti JP, Mayer JE Jr (2000) Ann Thorac Surg 70:140
282. Mallarde D, Valiere M, David C, Menet M, Guerin P (1998) Polymer 39:3387
283. Marois Y, Zhang Z, Vert M, Deng X, Lenz R, Guidoin R (1999) J Biomater Sci Polym Edn 10:483
284. Marois J, Zhang Z, Vert M, Deng X, Lenz R, Guidoin R (1999) J Biomat Sci Polym Edn 10:469
285. Shinoka T, Breuer CK, Tanel RE, Zund G, Miura T, Ma PX, Langer R, Vacanti JP, Mayer JE (1995) Ann Thorac Surg 60:S513
286. Stock UA, Wiederschain D, Kilroy SM, Shum-Tim D, Khalil PN, Vacanti JP, Mayer JE Jr, Moses MA (2001) J Cell Biochem 81:220
287. Sodian R, Sperling JS, Martin DP, Egozy A, Stock U, Mayer JE Jr, Vacanti JP (2000) Tissue Eng 6:183
288. Shinoka T, Shum-Tim D, Ma PX, Tanel RE, Isogai N, Langer R, Vacanti JP, Mayer JE (1998) J Thorac Cardiovasc Surg 115:536
289. McMillin CR (1994) Rubber Chem Technol 67:417

Adv Polym Sci (2006) 203: 63–93
DOI 10.1007/12_089
© Springer-Verlag Berlin Heidelberg 2006
Published online: 7 April 2006

Modulating Extracellular Matrix at Interfaces of Polymeric Materials

Carsten Werner (✉) · Tilo Pompe · Katrin Salchert

Leibniz Institute of Polymer Research Dresden, Max Bergmann Center of Biomaterials,
Hohe Str. 06, 01069 Dresden, Germany
werner@ipfdd.de

Abstract As extracellular matrices (ECM) closely interact with cells in living tissues and, through this, influence essentially any aspect of life *engineering of ECM* currently receives a lot of attention in the advent of regenerative therapies. Artificial matrices based on biopolymers isolated from nature were successfully utilized to prepare various types of cell scaffolds to enhance the integration and performance of engineered tissues. Beyond that, translation of progress in matrix and cell biology into new concepts of materials science permits to further refine the functional characteristics of such reconstituted matrices to direct tissue regeneration processes. The review emphasizes research to modulate the functionality of ECM biopolymers through their combination with synthetic polymeric materials. Two examples referring to our own studies concern (1) the control of vasculogenesis by adjusting the availability of surface bound fibronectin for cell-driven reorganization and; (2) the imitation of the bone marrow niche with respect to the cultural amplification of hematopoietic progenitor cells using collagen I-based assemblies. As a perspective we briefly discuss the design of biohybrid ECM mimics where synthetic and natural polymers are combined on the molecular scale for future use as morphogenetic templates in in vivo tissue engineering applications.

Keywords Extracellular matrix · Fibronectin · Collagen · Reconstitution

Abbreviations

2-D   two-dimensional
3-D   three-dimensional
BMP   bone morphogenetic protein
ECM   extracellular matrix
FGF   basic fibroblast growth factor
FN   fibronectin
MW   molecular weight
MMP   matrix metalloproteinase
PEG   poly(ethylene glycol)
PEMA   poly(ethylene-*alt*-maleic anhydride)
POMA   poly(octadecene-*alt*-maleic anhydride)
PPMA   poly(propene-*alt*-maleic anhydride)
RGD   arginine-glycine-aspartic acid
SDF-1   stromal cell-derived factor-1
VEGF   vascular endothelial growth factor

1
Introduction

The extracellular matrix (ECM) is a highly orchestrated ensemble of proteins, proteoglycans, and glycosaminoglycans arranged in tissue specific three-dimensional structures [1–4]. ECM plays a key role in tissue architecture and homeostasis by providing structural support and tensile strength, attachment sites for cell surface receptors, and feedback-controlled reservoirs for signalling factors related to a wide variety of processes related to cell differentiation, tissue formation, homeostasis, and regeneration. Thus, the "nonliving" material of the ECM biopolymers has to be considered to create the base for multicellular tissues and their dedicated functions in living organisms.

The principal proteinaceous components of the ECM are collagens which are secreted by a variety of stromal cells. Other proteins involved in different specialized matrix assemblies, such as the basement membrane [5], include fibronectin, laminin, collagen IV, and various growth factors and proteases. A second class of molecules that play an essential role in the ECM are proteoglycans whose protein core is covalently bound to high-molecular-weight glycosaminoglycans, including chondroitin and heparan sulfate. Proteoglycans support cell adhesion and bind various growth factors [6]. The ECM also contains hyaluronan, a glycosaminoglycan not linked to a core protein, which regulates adhesion, trafficking, and signalling of cells.

A survey of some of the most-abundant ECM biopolymers and their known functions is given in Table 1; for more details the reader is referred to [2].

Table 1 Most abundant biopolymers of the extracellular matrix

Proteins/glycoproteins	Glycosaminoglycans/proteoglycans
Collagens (wide variety of variants)	Chondroitin sulfate
Laminins	Heparan sulfate
Fibronectin	Keratan sulfate
Elastins (not glycosylated)	Hyaluronic acid (without protein component)

The decision of a cell to differentiate, proliferate, migrate, apoptose, or perform other specific functions is a coordinated response to the molecular interactions including ECM effectors. The flow of information between cells and their ECM is highly bidirectional involving ECM degradation, synthesis, and reorganization. In consequence, the ECM constantly undergoes changes in both structure and composition which range from dynamic homeostasis of resting-state adult organs to tissue remodelling, as occurs during development [7], inflammation [8], and wound healing [9].

Recently accumulated evidence emphasizes the importance of ECM-mediated interactions in morphogenesis. Lonai illustrated this view by examples of FGF signalling in vertebrate limb development [10]. The importance of ECM molecules in development has been further proven by in vivo studies using gene targeting. Mice lacking certain ECM component genes such as collagen, fibronectin, and laminin die before birth, others lacking component genes of molecules such as tenascin or osteonectin survive but exhibit characteristic defects [11, 12].

In front of this background it is hardly surprising that engineering extracellular matrix has been shown to be a most powerful means to influence cellular fate decisions involved in tissue regeneration and, therefore, recently received a lot of attention in the advent of regenerative therapies [13, 14]. Examples cover a wide range of approaches as different as adsorbing isolated matrix biopolymers to various cell culture carriers [15–19], molecular engineering of matrix components [20–23], and the design of multicomponent biopolymer assemblies providing morphogenetic cues and changing their characteristics during time and upon demand of embedded cells [24–27].

While a majority of earlier studies and applications used ECM biopolymers "as such", i.e. expecting a desired functionality from the presence of the molecules, it turned out to be of crucial importance *how* the matrix components are actually presented. In line with the occurrence of ECM in vivo [28], supramolecular association—in homo- as well as heteromolecular assemblies—was convincingly shown to create substantially different functional modes of ECM when reconstituted in vitro [29]. The varied accessibility of contained sites for biospecific interactions, e.g. with cellular receptors

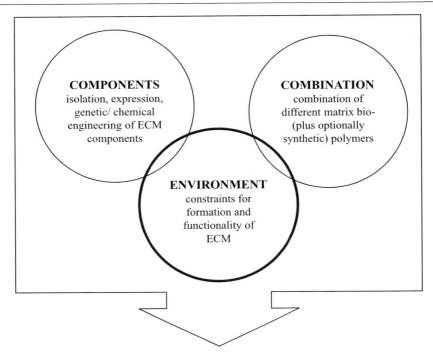

Scheme 1 Major aspects of the design of bio-artificial ECM mimics

or proteolytic enzymes, and of physical cues [30], such as pliability and contact guidance, define distinct states of ECM being highly relevant for cells to interact with. These characteristics can be controlled in vitro by the choice of matrix components to be combined, by their structural modification, and by their assembly including the formation of artificial chemical linkages. A second most powerful option to influence the features of bio-artificial ECM consists of the combination with synthetic materials. This includes both the provision of environmental constraints for pure biopolymer ECM assemblies and the formation of hybrid matrices combining synthetic and biomolecular structures.

This review places emphasis on options to modulate the functionality of ECM biopolymers through immobilization onto polymeric materials exhibiting a variety of physicochemical characteristics. Fundamental features of ECM will be introduced in a rather condensed fashion, followed by a survey on the current state of techniques to reconstitute ECM in vitro. In front of this background, we discuss options to switch biopolymer assembly and matrix functionality by immobilization onto polymer substrates and illustrate this part of the review with two examples referring to our recent studies. Finally, current trends towards more advanced bio-artificial matrices will be surveyed.

2
Biospecific and Physical Characteristics of Extracellular Matrices

ECM variants found in different tissues and the current knowledge of their structure-property relations are comprehensively summarized in several reviews, periodicals and monographs [1–9]. The crucial role of ECM in any of the most fundamental fate transitions of cells and in the development, maintenance and repair of tissues was demonstrated with a multitude of examples as summarized in [10].

The discussion of ECM often uses the categorization of biospecific and physical characteristics acting on the embedded cells. In that view biospecific properties are directly related to molecular entities and their signalling functions involved in cell adhesion, cell polarity, migration, proliferation, and differentiation. Physical characteristics give rise to functions exceeding the (inter)action of individual molecules governing cell scaffolding, the tensile strength of tissues, cushioning (as in cartilage), filter functions (as in the kidney), the formation of boundaries between, and within tissues, and the storage of growth factors, cytokines, and chemokines. The use of these categories involves arbitrariness as seen by the fact that physical characteristics of ECM often base on ensembles of biospecific bonds.

Biospecific signals of ECM concern the obvious role of the involved biopolymers to interact with molecular components of their environment in precisely defined ways. The biological responses of cells to the ECM are regulated by specific cell-surface receptors. Many different receptors have been identified that transduce signals including members of the heterodimeric integrins, receptor tyrosine kinases and phosphatases, immunoglobulin superfamily receptors, dystroglycan, and cell-surface proteoglycans. Most of the biologically active ECM molecules, including laminins, collagens, thrombospondin, and fibronectin, contain multiple active sites, often for different activities, and interact with different receptors. For example, some 40 active sites have been identified on laminin-1 and 20 different receptors have been characterized. In addition to that specific binding of ECM components includes the homo- and heteromolecular binding to other biopolymers of the ECM leading to the formation of supramolecular assemblies (e.g. the formation of fibrils of collagen I and of co-fibrils of collagen I glycosaminoglycans), the interaction with proteolytic enzymes, such as matrix metalloproteinases, which cause the degradation of matrix biopolymers through catalyzing the cleavage of certain bonds in the biopolymers' main chain, and the binding of growth and differentiation factors as well as chemokines modulating the presentation of theses molecules to the related cellular receptors via direct binding of the ECM-bound factors or through their localized release.

Proteolytic cleavage of extracellular matrix (ECM) proteins by matrix metalloproteinases and/or conformational changes unmask "cryptic" sites and liberate fragments with biological activities that are not observed in the un-

Table 2 Examples of interactions of some major ECM components via known binding sites

ECM component	Interaction/bound molecules
Collagen I	Self assembly, binding to integrin $\alpha_1\beta_1$ and $\alpha_2\beta_1$, chondroitin and dermatan sulfat, fibronectin, heparin
Collagen IV	Self assembly, binding to laminin, nidogen/entactin, perlecan, $\alpha_1\beta_1$ integrin
Laminins	Self assembly, nidogen/entactin, $\alpha_1\beta_1$ integrin, $\alpha_2\beta_1$ integrin, $\alpha_3\beta_1$ integrin, $\alpha_6\beta_1$ integrin, $\alpha_{17}\beta_1$ integrin, $\alpha_6\beta_4$ integrin
Fibronectin	Self assembly, fibrin, heparin sulfate, factor XIII, fibulin, $\alpha_5\beta_1$ integrin, $\alpha_4\beta_1$ integrin, $\alpha_{IIB}\beta_3$ integrin, $\alpha_V\beta_3$ integrin, collagen, other integrins
Heparan sulfate	Binding to a wide variety of growth factors including FGFs, BMPs, collagen I

modified molecule. The unmasking of cryptic sites is a tightly controlled process, reflecting the importance of cryptic ECM functions [31]. Exposure of binding sites through conformational changes of ECM proteins directing a sequential chain of binding steps was studied in detail for the formation of fibronectin–fibronectin bonds. Self-assembly sites exposed upon application of shear force on the protein when linked to integrin receptors was found to be a most efficient working principle for the advanced control of matrix reorganization and other processes [32]. Intriguing experimental and computational data are emerging to suggest that mechanical forces regulate the functional states of some proteins by stretching them into non-equilibrium states. Molecular design principles were reported to control the exposure of a protein's recognition sites, and/or their relative distances, in a force-dependent manner [33].

The formation of supramolecular assemblies of ECM biopolymers defines an obvious and highly important interconnection of physical and biospecific characteristics of the matrices. In general, multiple specific interactions between individual biopolymers, such as collagen or fibronectin, result in the formation of large fibrillar or meshwork-like structures segmenting tissues and guiding cells by microstructural features and mechanical characteristics. With those structures, often referred to as connective tissue, basement membrane, and several others, the unique functionality of ECM and the fact that it is based on both the provision of specific binding sites and physical forces is best represented.

Another aspect of ECM related to both biospecific and physical characteristics of ECM concerns the formation of gradients of any of the above-listed signals to direct cellular behavior. Such gradients are, for instance, key to the signalling of chemokines such as stromal cell-derived factor-1 (SDF-1) for the homing of circulating stem cells. Avigdor et al. [34] recently demonstrated that the adhesion receptor CD44 and its major ligand, hyaluronic acid, are essential for the homing of hematopoietic stem cells into the bone marrow and spleen of immunodeficient (NOD/SCID) mice. Stem cells migrating on hyaluronic acid toward a gradient of SDF-1 acquired a spread and polarized morphology with CD44 concentrating at the leading edge of the pseudopodia. Hyaluronic acid was concluded to play a key role in SDF-1-dependent transendothelial migration of hematopoietic stem cells and their final anchorage within specific niches of the bone marrow.

Likewise, spatial concentration gradients mediated by ECM biopolymers are the base of morphogens, signaling molecules that emanate from a restricted region of a tissue and spread away from their source [35]. Morphogens include members of the Hedgehog family, the Wnt family, and some members of the TGFß family; their gradients provide positional information to cells in the course of development. Recently, several reports have suggested that heparan sulfate proteoglycans determine morphogen transport and/or signaling. In particular, ECM seems to be critical in the restricted diffusion of morphogens in extracellular spaces, a most important mechanism by which morphogens traffic through tissues.

Physical cues of ECM include the spatial constraints and mechanical forces acting on cells from the environment.

A series of classical in vitro studies of Ingber and Whitesides [36–40] had shown that microstructuring a planar surface with an adhesive/anti-adhesive pattern established by the localized presence/absence of physisorption of matrix proteins was efficient to switch adherent endothelial cells from proliferation to apoptosis or vasculogenesis. In these experiments cell spreading was varied while maintaining the total cell-matrix contact area constant by changing the spacing between multiple focal adhesion-sized islands. The results revealed that cell shape govern whether individual cells grow or die, regardless of the type of matrix protein or antibody to integrin used to mediate adhesion. The mechanism behind these findings was suggested to concern the forces acting on the nucleus through the cytoskeleton [41, 42]. Gradual variations in cell shape distortion was shown to switch cells between distinct gene programs (e.g. growth, differentiation and apoptosis), and this process was discussed as a biological phase transition producing characteristic cell phenotypes.

Applying new concepts from nanotechnology the idea of lateral patterning was recently extended by Spatz et al. [43, 44] to clarify the impact of distances between adhesive ECM ligand structures on cell adhesion. To study the function behind the molecular arrangement of single integrins in cell adhesion hexagonally close-packed rigid templates of cell-adhesive gold nanodots were coated with cyclic RGDfK peptides by using block-copolymer micelle nanolithography. These dots were positioned with high precision at 28, 58, 73, and 85 nm spacing at interfaces. A separation of > or = 73 nm between the adhesive dots results in limited cell attachment and spreading, and dramatically reduced the formation of focal adhesion and actin stress fibres. The range between 58–73 nm was proposed to be a universal length scale for integrin clustering and activation, since these properties were found to be shared by a variety of cultured cells.

Furthermore, the pliability of the matrix providing adhesive ligands and the anchorage of the matrix to the environment was shown to determine the variant of cell-matrix adhesion developing by embedded cells which, in turn, may control their fate transitions (for an example see also Sect. 4.1). Cukierman et al. [45, 46] could convincingly show that based on a similar collagen-based network the 2D vs. 3D microstructure and their mechanical characteristics strongly influence the character of cell-matrix adhesions: The composition and function of adhesions in three-dimensional (3D) matrices derived from tissues or cell culture were found to differ substantially from focal and fibrillar adhesions characterized on 2D substrates in their content of $\alpha_5\beta_1$ and $\alpha_V\beta_3$ integrins, paxillin, other cytoskeletal components, and tyrosine phosphorylation of focal adhesion kinase. Relative to 2D substrates, 3D-matrix interactions were also reported to display enhanced activities and narrowed integrin usage. These distinctive in vivo 3D-matrix adhesions were demonstrated to differ in structure, localization, and function from classically described in vitro adhesions, and concluded to be more biologically relevant to living organisms.

Recently, the behavior of some cells on soft materials was found to be characteristic of important phenotypes. Discher et al. [47, 48] reported important facts to clarify how tissue cells—including fibroblasts, myocytes, neurons, and other cell types—sense matrix stiffness based on quantitative studies of cells adhering to gels (or to other cells) with which elasticity can be tuned to approximate that of tissues. Contractile myocytes were reported to sense their mechanical as well as molecular microenvironment, altering expression, organization, and/or morphology accordingly. The cells were cultured on collagen strips attached to glass or polymer gels of varied elasticity. Subsequent fusion into myotubes was found to occur independent of the substrate flexibility. However, myosin/actin striations emerged later only on gels with stiffness typical of normal muscle (passive Young's modulus approximately 12 kPa). On glass and on much softer or stiffer gels cells did not striate. Unlike sarcomere formation, adhesion strength was found to increase monotonically versus substrate stiffness with strongest adhesion on glass.

While the above-cited studies mainly revealed the impact of physical characteristics of ECM structures on cells grown in culture it seems justified to conclude that various kinds of microstructures occurring in natural matrix variants may act in similar ways. This holds particularly true for the guidance of cells with respect to growth in a certain orientation of larger ensembles as occurring in the formation—or conservation—of segmented structures in various tissues [49–52]. However, the wealth of in vivo data available on the relevance of biospecific interactions of ECM needs to be accomplished by further in vitro studies clarifying the impact of physical characteristics of ECM as well.

Altogether, ECM provides highly orchestrated signals which were identified for a wide variety of examples throughout the last years. While clearly more research is needed to unravel the cellular machineries responding to, reorganizing, deleting, and secreting ECM structures the challenge in the context of regenerative therapies is—already at the current premature stage of knowledge—to utilize ECM for the guidance of cells into tissue regeneration. A key aspect of this is the reconstitution of ECM assemblies in vitro.

3
Reconstitution of Extracellular Matrix Assemblies in vitro

As the functionality of ECM, in particular the characteristics of providing physical signals, is closely related to the self assembly of ECM into different types of supramolecular aggregates the reconstitution of these assemblies had to and did receive attention in the ongoing research towards engineered ECM.

The templates of matrix structures found in nature are often quite complex alloys or macromolecular composites and, so far, elucidated only for a number of cases. Also, interactions between different types of matrix assemblies have to be considered. For example, the most abundant matrix proteins, collagens, predominantly form either fibrillar or sheet-like structures—the two major supramolecular conformations that maintain tissue integrity. In connective tissues, other than cartilage, collagen fibrils are mainly composed of collagens I, III, and V at different molecular ratios, exhibiting a D-periodic banding pattern, with diameters ranging from 30 to 150 nm, that can form a coarse network in comparison with the fine meshwork of the basement membrane [53]. The basement membrane—also referred to as the lamina densa—represents a stable sheet-like meshwork composed of collagen IV, laminin, nidogen, and perlecan compartmentalizing tissue from one another. The interactions between collagen fibrils and the lamina densa seem to be mediated by collagen V and are considered to be crucial for maintaining tissue–tissue interactions [54].

In vitro reconstitution of ECM assemblies reported so far reached rather different degrees of faithfulness to the related type of matrix present in living tissues—while a close similarity of collagen I-based structures obtained upon reconstitution with those found in connective tissues the quality of reconstitution is often less perfect and hard to evaluate for other matrix types such as the basement membrane. A further type of assembly procedures may be distinguished that involves cell cultures (i.e. aided assembly as observed for fibronectin fibrillogenesis) from the spontaneous assembly of certain biopolymers such as collagen I. There is also a "top-down alternative" to the "bottom-up approach" of the in vitro reconstitution of ECM biopolymers which consists of the isolation of assembled matrix structures from living tissues without fragmentation, e.g. the collection of decellularized tissue as medically applied in the form of xenogenic (porcine) transplant heart valves.

Obviously, any "bottom up" reconstitution depends on the availability of isolated, dissociated ECM biopolymers. Enzymatically driven fragmentation of ECM forms like connective tissue often results in structurally altered biopolymers which may exhibit limited capability of reconstitution.

Collagen I-based fibrillar assemblies do certainly represent the best-studied example of in vitro reconstituted ECM structures [55]. Collagen is used as a generic term for proteins forming a characteristic triple helix of three polypeptide chains. So far, 27 genetically distinct collagen types have been described. Type I collagen forms more than 90% of the organic mass of bone and is the major collagen of tendons, skin, ligaments, cornea, and many interstitial connective tissues. The collagen type I triple helix is formed as a heterotrimer by two identical $\alpha1(I)$-chains and one $\alpha2(I)$-chain. The fibrils are indeterminate in length, insoluble, and form elaborate three-dimensional arrays that extend over numerous cell lengths. Studies of the molecular basis of collagen fibrillogenesis have provided insight into the trafficking of procollagen (the precursor of collagen) through the cellular secretory pathway, the conversion of procollagen to collagen by the procollagen metalloproteinases, and the directional deposition of fibrils involving the plasma membrane and late secretory pathway [56]. Fibrils arranged in elaborate three-dimensional arrays, such as parallel bundles, orthogonal lattices and concentric weaves provide the base for dedicated matrix structures in tendons and ligaments, in the cornea, and in bone. The fibrils are synthesized and secreted by fibroblasts and might self assemble thereafter but accumulating evidence suggests that fibril assembly can begin in the secretory pathway and at the plasma membrane.

The reconstitution of collagen I is attractive both for the relevance due to the abundance and dominating role in various tissues and for the relatively easy access and reconstitution. Enzymatically fragmented tropocollagen, collected from various kinds of connective tissue, can be dissolved in acidic solutions of sufficiently high ionic strength and precipitated into the fibril form upon neutralizing the solution. Variation of temperature and electrolyte concentration was demonstrated to influence the fibrillogenesis with respect to the dynamics of the process, however, the resulting fibrils show a very similar structure almost independent of the "history" of their formation and so do the rheological characteristics of the resulting gels. The spontaneous self-assembly of monomeric collagen was attributed to the entropy gain upon binding of collagen molecules implying that hydrophobic interactions between collagen monomers are the major driving force for fibril formation [1, 57]. In addition to hydrophobic interactions the contribution of water clusters bridging recognition sites on opposing helices was discussed as a further driving force for the fibril assembly [58]. Furthermore, recognition sites for the non-helical telopeptides of a docking collagen monomer are crucial for fibril formation and the lack of telopeptides increases the assembly time [59]. Mechanistically, fibrillogenesis was described as a multistep process initiated by the formation of collagen dimers and trimers followed by a rapid lateral aggregation involving five trimers [60]. Consecutive linear and lateral addition of further monomers or multimers gives rise to the formation of microfibrils which can further assemble into large fibrillar structures depending on the conditions of the fibrillogenesis [61]. In contrast to the highly ordered ensembles of collagen fibrils observed in the above-mentioned tissues reconstituted collagen I gels contain stochastically oriented fibrils only. However, application of shear force as well as electrical fields may permit us in future to create higher degrees of order in reconstituted structures as well.

Several attempts were reported to implement additional matrix biopolymers, namely glycosaminoglycans such as chondroitin sulfate, in collagen I-based fibrillar assemblies [62, 63] in order to obtain matrix mimics resembling more closely their template structure (see also Sect. 4.2 of this review). The mechanism of interaction was suggested to be dominated by electrostatic attraction between positively charged domains of collagen and the negatively charged polysaccharides [64]. Results of a recent own study point at the interfering effects of contact mediation and contact inhibition caused by heparin and hyaluronic acid during collagen I fibrillogenesis which may explain the drastic variation of the obtained fibril structures from the unique dimensions of the pure collagen I fibrils and in dependence on the contact conditions [65].

As an easily injectable, biocompatible matrix collagen gels are highly attractive for tissue engineering applications since they can act as a "cage" to retain cells or as gene delivery complexes [66]. In particular, bone tissue engineering relies on collagen I-based structures which resemble the exact compact bone matrix architecture over distances reaching centimeters and more [67]. To alter the mechanical properties and the cellular reorganization of collagen matrices for cell culture the influence of glutaraldehyde as a crosslinking agent was examined [68]. Even advanced strategies of regenerative therapy, such as the vasculogenesis of endothelial cells derived, purified and expanded in vitro from embryonic stem cells was recently shown to be supported by collagen I gels [69].

Unlike most collagens type IV collagen invariably forms an amorphous covalently stabilized network that is predominantly found in basement membranes. In addition to collagen type IV various other extracellular components, among them laminins, perlecan, nidogen, and fibulins were identified to contribute to the thin sheets of basement membranes which mechanically support tissue architecture, separate connective tissues from epithelial, endothelial, and muscle cells as well as provide immobilized ligands for cell surface receptors [70]. The supramolecular organization of collagen type IV is facilitated by different association sites of the collagen monomers resulting in the formation of dimers, tetramers, and lateral supertwisted associations stabilized by covalent bonds [71]. However, the proposed meshwork model of collagen IV [72] cannot explain the fine network of the basement membrane in tissues. Consequently, the concept of enmeshed laminin and collagen type IV polymers is the currently accepted model [73]. Various studies have shown that collagen type IV reassembles under physiological conditions in vitro. Interaction of laminin and collagen IV, both isolated from the Engelbreth-Holm-Swarm tumor, under physiological conditions were found to result in the precipitation of the basement membrane molecules while collagen alone gave no precipitate indicating that the components of the basement membrane interact in a highly specific manner [74]. However, later studies revealed that collagen type IV isolated from bovine lens capsules and from human placenta can construct the skeletal meshwork observed in the lamina densa under physiological conditions without addition of other macromolecules [75]. Interestingly, the reconstituted collagen type IV aggregates actually provided dimensionally similar geometries in comparison to collagen IV lattices which were seen in the lamina densa of the mouse pancreas.

Collagen IV/laminin matrices are widely used in cell culture studies. For example, phenotypic changes of human umbilical endothelial vein cells were analyzed with respect to the formation of a network of tubular structures with intercellular or lumen-like spaces. Ultrastructural analyses of the capillary-like structures and the mechanism of lumen-like formation indicated that the in vivo angiogenesis was better reproduced in the collagen IV/laminin model matrix as compared to the addition of FGF to the culture medium [76]. In vivo, these gels are used for measuring angiogenic inhibitors and stimulators, to improve graft survival and to repair various damaged tissues [77]. For example, liquid bioartificial tissue substitutes based on collagen IV/laminin containing embryonic stem cells were furthermore reported to constitute a powerful new approach to restoring injured heart muscle without distorting its geometry and structure [78].

The formation of fibronectin fibrils, another relatively well-studied supramolecular structure of ECM polymers, was extensively studied in vitro as well. Many different cell types synthesize fibronectin and secrete it as a disulfide-bonded dimer composed of 230–270 kDa subunits. Each subunit contains three types of repeating modules, types I, II, and III. These modules comprise functional domains that mediate interactions with other ECM components, with cell surface receptors and with fibronectin itself. Fibronectin matrix assembly involves binding domains and repeating modules from all regions of fibronectin which participate in interactions with cell surface receptors and with other fibronectin molecules. Newly secreted fibronectin dimers bind to transmembrane integrin receptors. Integrin $\alpha_5\beta_1$ is the primary receptor for mediating assembly linking fibronectin to the actin cytoskeleton. On the outside of the cell, interactions between fibronectin and integrins promote fibronectin–fibronectin association and fibril formation perhaps by inducing conformational changes in bound fibronectin. The thickness of fibronectin fibrils varies substantially from 10 to 1000 nm in diameter, extended fibronectin molecules are about 3 nm in diameter indicating that fibrils probably range from a few to several hundred fibronectin molecules across. The organization of fibronectin within the fibrils is not well understood but is probably determined by intermolecular interactions involving some subset of the following domains [79]. Detailed hypotheses on the mechanism of the fibronectin fibrillogenesis were derived in conceptually and experimentally elegant studies by Vogel et al. [39]. Part of this work was also the proof that shear force—applied upon expansion of a fibronectin monolayer at the air–water interface using a Langmuir trough–could be sufficient to create fibronectin–fibronectin interconnections, even without the protein linked to cellular receptors. Shear stress was also applied to fibronectin solutions in work done by Brown et al. [80] to produce fibronectin mats, a macroscopic fibronectin-based structure, which had been successfully used as culture carriers. The materials were prepared from human plasma fibronectin by applying fluid shear forces directly to a viscous solution of fibronectin. Structural analysis revealed that mechanical shear resulted in the formation of an orientated fibrous protein material that was less soluble than its non-sheared counterpart. Addition of fibrinogen and further modifications achieved by the processing [81] were investigated with respect to the structure and cell guidance characteristics of the fibronectin cables.

Recent reports suggest that fibronectin-rich matrices may play an important role in the formation of specialized matrices—the accumulation of fibronectin fibrils may be seen to contribute a transient auxiliary functionality as, for instance, discussed for the process of angiogenesis [82].

Fibrin, the polymerized form of the coagulation plasma protein fibrinogen, is, together with fibronectin deposited into wounds from the circulation shortly after injury and influences the reorganization of damaged tissues as a provisional matrix. Fibronectin deposition is continued by wound fibroblasts and macrophages, or migrating keratinocytes [83]. In response to injury, resident fibroblasts in the surrounding tissue proliferate and then migrate into the wounded site [84]. Once within the wound, fibroblasts produce type I procollagen as well as other matrix molecules and deposit these extracellular matrix molecules in the local milieu. Fibroblasts can use a fibrin and fibronectin matrix to move through the wound. When exposed to a chemotactic gradient, they will migrate along, rather than across, the fibronectin fibrils. Fibrin plays an important role in healing and regeneration in the developed organism, but not in embryonic and fetal development. As a result of this, few of the interesting morphogenetic signals involved in development interact with fibrin in a specific manner. However, fibrin glue is widely used for wound closure and has been shown to be a suitable delivery vehicle for exogenous growth factors that may in the future be used to accelerate wound healing [85]. Although the regenerative capacity of the organism is not activated in pure fibrin matrices the addition of heparin or engineered adhesion and growth factors incorporated into the fibrin matrix during coagulation has been thoroughly studied as a promising means to modulate the morphogenetic characteristics of this temporary matrix. Thus, fibrin received a lot of attention in the design of engineered or biohybrid matrices distinguished by a rather high rate of turnover and a readily tuneable variety of characteristics.

Altogether, a variety of ECM assemblies has been structurally elucidated and some of those became available through in vitro reconstitution. Cell-free reconstitution provided a wealth of supramolecular ECM mimics which, however, often base on one enzymatically processed biopolymer component— such as collagen I—only while others may deviate in structure and composition in a unknown fashion from their natural templates (as, for instance the basement membrane precipitates from laminin and collagen IV isolated from the Engelbreth-Holm-Swarm tumor). Although successfully used in different cell culture experiments, the model characteristics of biopolymer assemblies obtained via cell-free reconstitution are limited with respect to ECM occurring in living matter. The use of cell-driven ECM reconstitution in vitro, in particular the formation of cell-secreted multicomponent assemblies, and the detailed comparison of structures resulting from both cell-driven and cell-free reconstitution with those present in living organisms may certainly open new options to further advance the model characteristics of ECM assemblies.

4
Matrix Assembly and Functionality at Polymer Surfaces

Having considered the formation of supramolecular associations between ECM biopolymers in vitro the question of factors influencing this process in artificial settings is emerging. While the choice of the biopolymers to be combined as well as the conditions of their precipitation (absolute and relative concentrations, solution pH and electrolyte composition, temperature, shear force, electrical fields and other physical factors) obviously trigger ECM reconstitution another, often underestimated aspect, concerns the presence of solid surfaces interacting with the ECM biopolymer prior to and/or during assembly.

Several recent examples demonstrate the importance of the modulation of ECM assemblies by their interaction with the surfaces of solid substrates. As hydrophobic dehydration and electrostatic forces certainly provide the most important contributions to protein adsorption as well as to the formation of supramolecular structures in aqueous environments the related surface characteristics of the solids were considered most carefully. Emphasis was put on the comparison of hydrophobic and hydrophilic surfaces used as substrates for the precipitation of protein assemblies and various studies used atomic force microscopy (AFM) to unravel differences in the interfacial protein structures. For example, it was reported that on hydrophilic mica β-amyloid peptides formed particulate, pseudomicellar aggregates, but on hydrophobic graphite the same protein organized into uniform, elongated sheets [86]. In another study, recombinant human elastin peptides were shown to adsorb as discrete, rounded aggregates on hydrophilic surfaces but on a hydrophobic surface the peptides self-assembled into an energetically favorable hexagonally closed packed fibril arrangement [87]. Also, for the binding of integrins to surface-bound fibronectin substrate-specific conformational changes have been inferred on hydrophobic and hydrophilic surfaces [88] and correlated with solid-phase cell binding assays [89]. Providing another instructive example Sherratt et al. [90] demonstrated that fibrillin and type VI collagen microfibrils exhibit substrate dependent morphologies. Recent work of Muller and coworkers [91] demonstrated how well-defined artificial assemblies can be obtained using the atomic force microscope as a biomolecular manipulation machine. Native collagen I molecules were mechanically directed on mica surfaces into well-defined, two-dimensional templates exhibiting patterns with feature sizes ranging from a few nanometers to several hundreds of micrometers.

As seen from these examples, insights on protein adsorption gained in colloid and interface science throughout the previous years [92, 93] are highly relevant for the understanding of ECM brought into contact with artificial materials. The formation as well as the functionality of the resulting matrix structures is strongly influenced by environmental constraints. This can be utilized not only to more efficiently mimic natural ECM templates, i.e. tissue-specific matrices, by artificially reconstituted matrix structures but also to design matrices which deviate in structure and function from any natural ECM to (re)activate pathways of tissue regeneration absent or insufficient in the adult organism. Expanding this idea the novel discipline of "matrix engineering" can be expected to play a key role in the future perspective of regenerative medicine.

As polymer materials permit a most versatile variety of surface characteristics efficient control over processes of ECM reconstitution can be achieved by the interaction of polymeric materials with the biopolymers of the ECM. With the following subsections the modulation of ECM biopolymer assembly at polymer interfaces is illustrated for two selected examples of own research in more depth and the relevance of the resulting differences is discussed for cellular systems.

4.1
Fibronectin Anchorage to Polymer Films Directs Fibrillogenesis
by and Angiogenesis of Endothelial Cells

This subsection is intended to illustrate the impact of a variation of the anchorage strength of fibronectin to polymer substrates on the formation of fibronectin fibrils and its relation to the morphogenesis of vascular-like structures of endothelial cells. The physicochemical surface characteristics of the substrates is modulated by a polymeric model system consisting of thin films of maleic anhydride copolymers. The used copolymers can be covalently attached onto a variety of substrates bearing amine moieties due to the high reactivity of the anhydride functionality towards primary amines. The versatility and broad applicability of this platform technology was recently described in more detail in [94] for example to modify silanized glass surfaces and silicon wafers, and low pressure plasma functionalized polymers. One interesting feature of this model system is the possible combinatorial approach to protein attachment. The different chemical attachment schemes of covalent binding of proteins to anhydride bearing surfaces or physisorptive anchorage to hydrolyzed surfaces bearing carboxylic acid groups are combined with a variability of the kind of the comonomer leading to a broad variation in the anchorage strength of the protein to the substrates. Although the surface chemistry remains almost constant—despite the variation of the surface density of carboxylic acid groups—the protein–substrate interaction can be varied between covalent and non-covalent binding as well as different degrees of polar and hydrophobic interactions.

The impact of different maleic anhydride copolymer surfaces on the adsorption behavior of fibronectin was investigated in detail by protein adsorption, and exchange experiments in [95, 96] and revealed dominant differences in the fibronectin anchorage strength besides slight conformational changes in the tertiary protein structure. The variation of fibronectin anchorage strength was measured in terms of the displacement kinetics by human serum albumin, and characterized by a double exponential decay in fibronectin surface coverage. Figure 1 shows the dependence of the fast displaced species on three different maleic anhydride copolymers, i.e. poly(octadecene-*alt*-maleic anhydride), poly(propene-*alt*-maleic anhydride), poly(ethylene-*alt*-maleic anhydride).

On the basis of these data the impact of a varied protein anchorage strength on fibronectin fibrillogenesis by endothelial cell was investigated by fluorescence and scanning force microscopy. Initial investigations [97] revealed a clear dependence of focal adhesion and fibronectin fibril formation. While covalent fibronectin attachment resulted in an impaired development of focal adhesions and fibronectin fibrils, an enhanced fibrillogenesis was observed on more hydrophilic substrates with physisorptive protein anchorage. More refined experiments [98] allowed for a quantification of the fibronectin

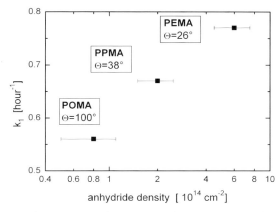

Fig. 1 Fibronectin anchorage strength in terms of time constant of double exponential fibronectin displacement kinetics ($\Gamma = \Gamma_1 \exp[- k_1 t] + \Gamma_2 \exp[- k_2 t]$) on different maleic anhydride copolymer surfaces characterized by the density of anhydride functionalities and water contact angle (poly(octadecene-*alt*-maleic anhydride)—POMA, poly(propene-*alt*-maleic anhydride)—PPMA, poly(ethylene-*alt*-maleic anhydride)—PEMA)

fibril pattern together with the focal adhesion density and a correlation of these features to the variation in fibronectin substrate anchorage. As shown in Fig. 2 endothelial cells can reorganize rhodamine-conjugated fibronectin to a much greater extent on the hydrophilic poly(ethylene-*alt*-maleic anhydride) substrate. Furthermore, the mean distance between the fibronectin fibrils was found to be smaller on those substrates. Together with the analysis on other copolymer substrates the fibril spacing could be directly correlated to the fibronectin anchorage strength—characterized by the time constant of fibronectin heteroexchange—as is shown in Fig. 3.

Together with an analysis of the focal adhesion pattern the following working model could be established from these results demonstrating the impact of the substrate physicochemistry on fibronectin fibrillogenesis. The overall

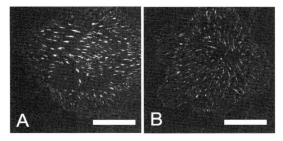

Fig. 2 Pattern of fibronectin fibrils after 50 minutes of reorganization by endothelial cells on poly(octadecene-*alt*-maleic anhydride) (**A**) and poly(ethylene-*alt*-maleic anhydride) (**B**). *Scale bar*: 20 µm

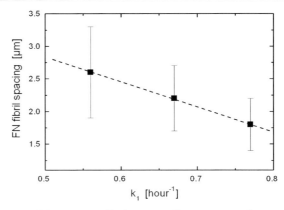

Fig. 3 Mean spacing of fibronectin fibrils versus fibronectin anchorage strength on different maleic anhydride copolymer substrates in terms of time constant of fibronectin heteroexchange

applied tension of the cytoskeleton towards the substrate can be assumed with a similar strength because the cells adhere and spread similarly on the different copolymer surfaces. As a different focal adhesion density was observed for the different substrates, the local force at those adhesion sites should scale indirectly proportional to their density, because focal adhesions are thought of as primary adhesion sites. The observed dependence of focal adhesion density on the substrate physicochemistry can be interpreted as the result of the cellular signalling mechanisms, which regulates the force applied on the adhesion ligands [99, 100]. Because of the different anchorage strength of fibronectin to the substrates, the cells establish a local force on the receptor-ligand pairs appropriate to the ligand-substrate anchorage strength. Subsequently, the focal adhesion pattern will function as a template for fibronectin fibrillogenesis, because the focal adhesions act as primary sites of fibronectin fibril formation [101]. Thus, the overall process results in a fibronectin fibril pattern with a spacing depending on the fibronectin substrate anchorage strength—directly influenced by the substrate physicochemistry.

Additional investigations [102] on the nanoscale fibronectin fibril pattern supported the hypothesis of cellular sensing of the fibronectin anchorage strength. Therein, a substrate-dependent spacing of paired fibronectin nanofibrils was correlated to the force sensitive feedback mechanism of the actin cytoskeleton and its inner structure. A larger spacing of the paired nanofibrils in the range of 300 nm was observed on substrates with a higher fibronectin anchorage strength. The distinct spacings were concluded to reflect the thickness of the actin stress fibres as template structures of fibronectin fibrillogenesis. The inner structure of actin stress fibres with α-actinin cross-linked actin filaments lead to a repeating unit of 71 nm for the spacing of the paired nanofibrils on the different copolymer substrates down to a minimum of one repeating unit on the most hydrophilic polymer [poly(ethylene-*alt*-maleic anhydride)], e.g. in Fig. 4. This again correlates very well with the model of cellular sensing of the fibronectin anchorage strength to the substrate and the corresponding intracellular feedback of force and focal adhesion size regulation.

As already mentioned, downstream cellular processes like proliferation and differentiation can be affected by extracellular signals, too. In order to address this issue the behavior of endothelial cells was investigated not only in respect to fibronectin fibrillogenesis on short-time scales. The proliferation and differentiation of endothelial cells into vascular-like tubes was analyzed in cell culture experiments over 5 days on substrates with a different fibronectin anchorage strength. As described in detail elsewhere [103], endothelial cells grew in a flat monolayer after 1 to 3 days of cell culture on all substrates. However, after 5 days distinct morphological differences were observed. On substrates with a high fibronectin anchorage strength endothelial cells still exhibited monolayer characteristics. In contrast, an early vascular-like network formation was observed on the substrate with a weaker fibronectin anchorage strength. The cells also started to form tubular structures. As an underlying reason for the different morphology, the formation of distinctively different extracellular matrix networks was found as shown in Fig. 5. The cell-derived fibronectin network on substrates with a high fibronectin anchorage strength showed a dense and fine structure and was restricted to the substrate surface below the cells. The formation of a vascular-like structure was accompanied by the formation of a coarse fibronectin network which was distributed around the cellular structures also in the vertical direction. The analysis by confocal laser scanning microscopy with imaging in different horizontal planes allowed us to quantify the differential matrix distribution (Fig. 5).

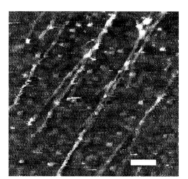

Fig. 4 Topography image of fibronectin nanofibrils after 50 min of reorganization by endothelial cells on poly(ethylene-*alt*-maleic anhydride) substrates visualized by scanning force microscopy at physiological conditions. The height of fibrils is 5 to 10 nm. *Scale bar*: 300 nm

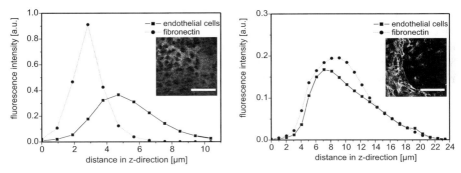

Fig. 5 Vertical distribution of the fibrillar fibronectin network (stained with fluorochrome conjugated antibodies) around endothelial cells (stained with CellTrackerGreen (Invitrogen)) after 5 days of cell culture on two polymer substrates with different anchorage strength [*left*—poly(octadecene-*alt*-maleic anhydride), *right*—poly(ethylene-*alt*-maleic anhydride)]. The *insets* show the lateral network pattern (*scale bar*: 100 μm). Visualization and analysis was performed on a confocal laser scanning microscope (TCS SP1, Leica)

From the reported findings one can conclude that gradated physicochemical characteristics of polymer substrates can be used to control the anchorage strength of extracellular matrix proteins. By the force sensitive cell surface receptors and the intracellular signalling mechanism the modified matrix characteristics subsequently affects not only cell adhesion and matrix reorganization, but furthermore cellular development in terms of differentiation. In a more general sense these findings can nicely be discussed in the concept of the "tensegrity" model [104], which combines extracellular signals like mechanical and structural features with intracellular processes like proliferation and differentiation.

4.2
Collagen Fibrils and Collagen-Glycosaminoglycan Cofibrils on Planar Surfaces and in 3D Carrier Materials to Imitate the Hematopoietic Niche

Various processing techniques have accompanied the reconstitution of collagen I-based cell scaffolds. In particular, chemical crosslinking or irradiation were widely applied not only to enhance the mechanical properties of the resulting materials but also to alter the degradation characteristics towards slower changes [105, 106]. As collagen type I specifically interacts with nearly 50 ligands, among them other extracellular matrix components, and growth and differentiation factors [107] another option consists of the in vitro decoration of collagen type I fibrils with different biopolymers such as fibronectin, vitronectin, or glycosaminoglycans such as heparin or chondroitin sulfate without chemical or physical activation of the collagen.

To provide tissue-mimetic environments for adherent cells glycosamino-glycans such as heparin and hyaluronic acid were implemented into three-dimensional collagen gels during self-assembly of monomeric collagen [8]. Turbidity measurements were utilized to follow the formation of collagen fibrils in the presence of heparin and hyaluronic acid that was initiated by an increase in temperature, pH value, and electrolyte content of a cold acidic solution of collagen monomers in the bulk volume. Varied optical densities were observed for different concentrations of glycosaminoglycans in comparison to pure collagen (Fig. 6). Gradually increasing portions of heparin and hyaluronic acid, respectively, at constant collagen concentrations caused a slight decline in the maximum optical densities indicating that fibril formation was obviously affected by the presence of the glycosaminoglycans. The differences in turbidity values were concluded to be either caused by different quantities of collagen fibrils or varying fibril diameters.

To covalently immobilize collagen and its assemblies with glycosamino-glycans, fibrillogenesis was performed in the presence of polymer-coated substrates resulting in thin layers of collagen fibrils. Therefore, cold mixtures of dissolved collagen and heparin or hyaluronic acid at different concentrations were exposed to glass slides or silicon wafers which had been modified before with thin films of poly(octadecen-*alt*-maleic anhydride) followed by the initiation of fibrillogenesis through an increase in temperature. The re-

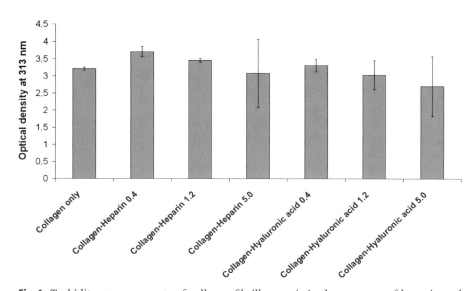

Fig. 6 Turbidity measurements of collagen fibrillogenesis in the presence of heparin and hyaluronic acid 2 h after initiation of fibril formation. Initial concentration of the non-fibrillar collagen solution was 1.2 mg/ml. Concentrations of the glycosaminoglycans were 0.4, 1.2, and 4.0 mg/ml, respectively

sulting collagen layers were characterized with respect to dry layer thickness utilizing ellipsometry, immobilized protein quantities using acidic hydrolysis of collagen and subsequent amino acid-based HPLC-analysis [108], and assessment of surface topography by scanning force microscopy and scanning electron microscopy. Furthermore, confocal laser scanning microscopy was used for the detection of surface-bound collagen-glycosaminoglycan assemblies. The data obtained revealed differences in the deposited collagen layers including thickness, protein amount, and surface character.

While pure collagen provided a dry layer thickness of about 80 nm, significantly thinner layers in the range between 14 and 32 nm (for fixed, pre-set refractive index values) were measured if heparin or hyaluronic acid were added to fibril-forming solutions (Table 3). The quantification of surface-bound proteins confirmed the results of the ellipsometric measurements: noticeable lower collagen amounts were quantified if fibril formation and immobilization was performed in the presence of glycosaminoglycans. The quantification was supported by surface topographic analysis of the attached fibrillar components (Fig. 7).

In comparison to the meshwork formed by pure collagen different surface patterns concerning shape and dimensions of the deposited fibrils depending on the glycosaminoglycans could be demonstrated. Collagen fibrils which had been reconstituted in the presence of polymer-coated substrates provided well-established networks without gaps between the separate fibres whereas fibril formation by addition of heparin or hyaluronic acid gave rise to modified morphologies of the deposited layers. In the case of heparin cofibrils

Table 3 Deposition of collagen-glycosaminoglycan conjugates on poly(octadecen-*alt*-maleic anhydride)-coated substrates. Fibrillogenesis and immobilization of fibrillar collagen (1.2 mg/ml) was performed for 2 h at 37 °C in the presence of heparin and hyaluronic acid (0.4, 1.2 and 5.0 mg/ml, respectively). Thickness of the layers was determined ellipsometrically using the refractive index of 1.6035 for the dried collagen layer. The collagen amount was determined by amino acid-based HPLC analysis after acidic hydrolysis of surface-bound collagen

Immobilized assemblies	Layer thickness [nm]	Collagen amount [$\mu g/cm^2$]
Collagen	81.3 ± 5.8	10.4 ± 0.4
Collagen/Heparin 0.4 mg/ml	n.d.	2.7 ± 1.2
Collagen/Heparin 1.2 mg/ml	29.3 ± 8.5	2.3 ± 0.3
Collagen/Heparin 5.0 mg/ml	32.3 ± 1.6	2.3 ± 0.9
Collagen/Hyaluronic acid 0.4 mg/ml	n.d.	3.4 ± 1.3
Collagen/Hyaluronic acid 1.2 mg/ml	14.8 ± 2.0	1.8 ± 0.5
Collagen/Hyaluronic acid 5.0 mg/ml	30.9 ± 8.7	3.0 ± 0.8

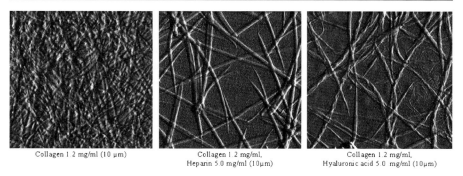

Collagen 1 2 mg/ml (10 µm)   Collagen 1 2 mg/ml, Hepann 5 0 mg/ml (10µm)   Collagen 1 2 mg/ml, Hyaluronic acid 5 0 mg/ml (10µm)

Fig. 7 AFM images of collagen and collagen-glycosaminoglycan fibrils attached to thin films of poly(octadecen-*alt*-maleic anhydride). Image size is given in brackets

shape and dimensions of the surface-bound fibrils significantly differed from those of pure collagen. Wide and straight fibrils were formed in the presence of heparin whereas hyaluronic acid caused the generation of more twisted collagen assemblies. In both cases no dense fibrillar network was formed, only separate fibrils were attached to the polymer surface and many small fibrils filled the spacings between the larger filaments. Additionally, the deposited layers appeared more smooth and thinner in comparison to pure collagen meshworks. All fibrillar variants revealed their characteristic banding pattern indicating that the native structure was retained after deposition. These observations endorsed the ellipsometric measurements as well as the results of quantified surface-bound collagen.

It should be noticed that the surface pattern of the immobilized collagen-glycosaminoglycan assemblies does not reflect the volume phase composition of collagen gels reconstituted in the presence of heparin or hyaluronic acid. When desiccating such gels on top of planar surfaces quite different surface structures regarding fibril size and density were observed. Larger and more elongated fibrils forming a tight fibrillar meshwork were detected in that case. Thus, fibrils attached to the polymer-coated surfaces did not represent the meshwork features of the gel bulk phase. The differences between surface-bound fibrillar collagen and collagen fibrils which were deposited in the presence of glycosaminoglycans were attributed to strong interactions of collagen with heparin or hyaluronic acid, respectively. Obviously, the attachment of the highly negatively charged glycosaminoglycans to the protein collagen— caused by ionic interactions as described for chondroitin-6-sulfate [2] or heparin [7] and via specific binding sites as specified for heparin [109]—may partially occupy the positively charged lysine side chains resulting in a reduced attachment of collagen to the poly(octadecen-*alt*-maleic anhydride) films. Furthermore, glycosaminoglycans most probably affect the cohesion of the fibrillar meshworks and therefore less fibrils remained on the polymer-coated surface after displacement of the gel volume phase.

Fluorescent-labeled collagen, heparin, and hyaluronic acid were used for the visualization of interactions between collagen and glycosaminoglycans attached to polymer films by confocal laser scanning microscopy (Fig. 8). Compared to the fluorescent-labeled layer of pure collagen, heparin induced the formation of straight fibrillar structures whereas hyaluronic acid containing assemblies provided less sharp structures. The detection of fluorescence in close proximity to collagen fibrils demonstrated the strong interactions between collagen fibrils and glycosaminoglycans. Thus, the results permit us to conclude on the incorporation of heparin or hyaluronic acid into fibrillar collagen structures.

To obtain a first measure for the interaction of cells with fibrillar and non-fibrillar matrix coatings in vitro the migration of human hematopoietic stem cells on various collagen substrates was analyzed [110]. Therefore, hematopoietic stem cells were isolated from umbilical cord blood using immunomagnetic selection and subsequently cultivated on collagen and collagen-glycosaminoglycan-modified cell culture carriers for 4 days. After removal of non-adherent cells migration of the remaining cell fraction was followed using time-lapse microscopy (Fig. 9).

In comparison to pure non-fibrillar tropocollagen minor migration rates were measured on fibrillar collagen indicating a stronger interaction of the cells with the fibrillar structures. Interestingly, the migration rates on collagen-glycosaminoglycan assemblies were again slower compared with fibrillar collagen. One possible explanation for the reduced motility of the cells on substrates containing heparin could be the binding of heparin/heparan sulfate receptors of hematopoietic stem cells to the collagen-bound heparin which was similarly described for endothelial cell surface heparin sulfates which participate in the adhesion cascade between hematopoietic progenitor cells and bone marrow endothelial cells and thus support the homing of hematopoietic progenitor cells [111]. The low cell migration on substrates containing hyaluronic acid could be ascribed to the high expression of CD44, the receptor for hyaluronic acid, which also plays an important role in homing and mobilization of hematopoietic stem cells [112].

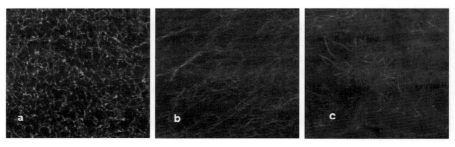

Fig. 8 Confocal laser scanning microscopy of collagen fibrils visualized by the addition of collagen-FITC (**a**), hyaluronic acid-FITC (**b**) and heparin-FITC (**c**). Image size is 125 μm

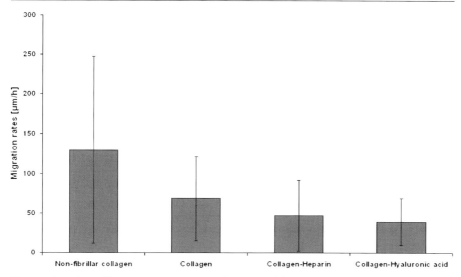

Fig. 9 Migration of hematopoietic stem cells isolated from umbilical cord blood on various collagen samples. Concentration was 1.2 mg/ml for collagen and 0.4 mg/ml for heparin and hyaluronic acid, respectively

Additional cell culture experiments were carried out on patterned 3D substrates in order to combine the spatial constraints of a stem cell niche with the influence of different extracellular matrix components of its microenvironment. Therefore, silicone molds of cavities with a size of 10 to 80 μm and a depth of 10 μm were coated with reactive maleic anhydride copolymers and subsequently reconstituted assemblies of collagen fibrils and cofibrils were immobilized on top of it. Hematopoietic stem cells were seeded onto these functionalized cell carriers. Depending on the size of the cavities cells differentially adhered to these structures as exemplarily shown in the inset of Fig. 10. In initial experiments adhesion and proliferation was followed over a time period of 3 days. As shown in Fig. 10 the cells tend to preferentially adhere to an intermediate size of cavities. This result points to a balanced equilibrium of cell-cell contacts and cell-matrix contacts for the homing and proliferation of hematopoietic stem cells, which is in agreement with the observed small hematopoietic stem cell clusters found in vivo [113]. Further experiments will follow up on these observations in order to unravel the influence of different extracellular matrix compositions combined with microcavities.

The introduced method for the preparation and in situ attachment of fibrillar collagen and glycosaminoglycans assemblies to reactive polymer coatings comprises an efficient technique for a straightforward biofunctionalization of cell culture carriers and medical devices. As a basic advantage the

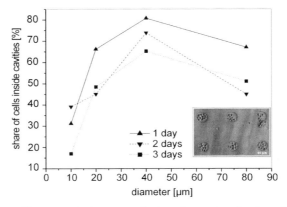

Fig. 10 Adhesion of hematopoietic stem cells to micrometer-sized cavities. The relative amount of cells inside the cavities in comparison to all cells on the surface after 1 to 3 days of cell culture

procedure allows a coating without chemical activation or irradiation and thus conserves the structural and functional features of the biopolymers and the underlying polymer substrate.

5
Perspective: Bio-Hybrid Extracellular Matrices

The previous section referred to systems in which ECM biopolymers interact with solid surfaces, in particular those of synthetic polymer materials, and could be modulated with respect to structure formation and functionality. Although the options of this approach deserve further attention the use of synthetic macromolecules in the engineering of ECM structures was recently shown to enable a level of interaction beyond this. Biohybrid polymer structures containing both synthetic and natural or synthetic and bio-analogous components have been introduced by the pioneering studies of a number of researchers. The rational of this concept concerns the functional characteristics of the resulting matrices, e.g. to stress certain desired functions more then natural or truly biomimetic matrices would allow, as well as several practical aspects related to processing and safety of the matrices. The term "biohybrid" is considered here to cover a range of rather different structures including chemically modified biopolymers (e.g. for crosslinking or decoration with active units) as well as synthetic hydrogels with bioactive (biomimetic) building blocks such as peptides triggering adhesion or enzymatic cleavage.

Jeffrey A. Hubbell pioneered the field of biomimetic polymer structures emphasizing the idea of responsiveness to cellular stimuli through the combination of a variety of incorporated bioactive building blocks [114]. He often utilized conjugate addition reactions of poly(ethylene glycol) (PEG) units and oligopeptides for the formation of polymer networks. For that purpose, vinyl sulfonate-functionalized multiarm telechelic PEG macromers were converted with the thiolate units of various cystein-containing peptides [115]. The creation of synthetic ECM analogs, in which ligand type, concentration and spatial distribution can be modulated upon a passive background was shown to help in deciphering the complexity of signaling in cell-ECM interactions. For instance, the suitability of proteolytically degradable synthetic poly(ethylene glycol) (PEG)-based hydrogels as an ECM model system for cell migration research was recently tested and compared with the two well-established ECM mimetics fibrin and collagen. The study shows that the high protease sensitivity makes PEG hydrogels an interesting model system that allows a direct correlation between protease activity and cell migration [116]. Also, the utility of integrating two types of protein signals into such networks has been demonstrated by combining oligopeptides mediating cell adhesion and substrates for cellular-controlled degradation to engineer bone [117]. Further relevant studies on the combination of peptides with synthetic polymer structures concern work to clarify the impact of the ligand density required for a particular cellular response [118], nanoscale spatial organization of adhesion ligands [119], the relevance of ligand gradients [120] and studies on the coregulation of signals [121]. Mooney and coworkers reported about a single matrix capable of delivering multiple growth factors (VEGF and PDGF) with distinct kinetics which dramatically increased blood vessel maturity [122].

While the above systems employ a synthetic polymer as the main component which is decorated with biomimetic peptides a rather different type of biohybrid matrix uses biopolymers as the main component which is further modified in order to exhibit certain desired characteristics. This approach has been elaborated in quite some detail for fibrin, the polymerized form of fibrinogen (see above). For example, an engineered variant form of VEGF that mimics matrix-binding and cell-mediated release by local cell-associated enzymatic activity was bound to fibrin with the result that the quality of angiogenesis in the matrix was substantially improved [123].

In view of these examples joint efforts of matrix biology and polymer science towards novel biohybrid matrices with tissue specific or morphogenetic signaling characteristics seem to be within in reach. Future work will have to elaborate this idea in the branches of regenerative therapies, namely stem cell bioengineering and in vivo tissue engineering, using the principles described above.

References

1. Wassarman P, Miner J (2005) Extracellular Matrix in development and disease: advances in developmental biology, vol 15. Elsevier, Amsterdam
2. Kreis T, Vale R (1993) Extracellular matrix and adhesion proteins. Oxford University Press, Oxford

3. Zern MA (ed) (1993) Extracellular matrix: chemistry, biology and pathobiology with emphasis on the liver. Marcel Dekker Ltd, Oxford
4. Kleinmann HK (1993) Extracellular matrix: advances in molecular and cell biology, vol 6. Elsevier, Amsterdam
5. Yurchenco PD, Amenta PS, Patton BL (2004) Matrix Biol 22:521
6. Maquart FX, Bellon G, Pasco S, Monboisse JC (2005) Biochimie 87:353
7. Kadler K (2004) Birth Defects Res C Embryo Today 72:1
8. Midwood KS, Williams LV, Schwarzbauer JE (2004) Int J Biochem Cell Biol 36:1031
9. Tran KT, Griffith L, Wells A (2004) Wound Repair Regen 12:262
10. Lonai P (2003) J Anat 202:43
11. Gustafsson E, Fassler R (2000) Exp Cell Res 261:52
12. Kleinman HK, Philp D, Hoffman MP (2003) Curr Opin Biotechnol 14:526
13. Lutolf MP, Hubbell JA (2005) Nat Biotechnol 23:47
14. Badylak SF (2005) Anat 287B:36
15. Barbucci R, Magnani A, Chiumiento A, Pasqui D, Cangioli I, Lamponi S (2005) Biomacromolecules 6:638
16. Dupont-Gillain ChC, Pamula E, Denis FA, De Cupere VM, Dufrene YF, Rouxhet PG (2004) J Mater Sci Mater Med 15:347
17. Dupont-Gillain ChC, Alaerts JA, Dewez JL, Rouxhet PG (2004) Biomed Mater Eng 14:281
18. Hori Y, Inoue S, Hirano Y, Tabata Y (2004) Tissue Eng 10:995
19. Healy KE, Rezania A, Stile RA (1999) Ann N Y Acad Sci 875:24
20. Kurihara H, Nagamune T (2005) J Biosci Bioeng 100:82
21. Cutler SM, Garcia AJ (2003) Biomaterials 24:1759
22. Richards CD, Kerr C, Tong L, Langdon C (2002) Biochem Soc Trans 30:107
23. Ogiwara K, Nagaoka M, Cho CS, Akaike T (2005) Biotechnol Lett 27:1633
24. Schense JC, Hubbell JA (2000) J Biol Chem 275:6813
25. Rizzi SC, Hubbell JA (2005) Biomacromolecules 6:1226
26. Pittier R, Sauthier F, Hubbell JA, Hall H (2005) J Neurobiol 63:1
27. Schmoekel HG, Weber FE, Schense JC, Gratz KW, Schawalder P, Hubbell JA (2005) Biotechnol Bioeng 89:253
28. Parry D, Squire M (eds) (2005) Fibrous proteins: coiled-coils, collagen and elastomers [Advances in protein chemistry (vol 70)]. Academic Press, New York
29. Wess TJ, Cairns DE (2005) J Synchrotron Radiat 12:751
30. Silver FH, Siperko LM (2003) Crit Rev Biomed Eng 31:255
31. Schenk S, Quaranta V (2003) Trends Cell Biol 13:366
32. Vogel V, Thomas WE, Craig DW, Krammer A, Baneyx G (2001) Trends Biotechnol 19:416
33. Krammer A, Craig D, Thomas WE, Schulten K, Vogel V (2002) Matrix Biol 21:139
34. Avigdor A, Goichberg P, Shivtiel S, Dar A, Peled A, Samira S, Kollet O, Hershkoviz R, Alon R, Hardan I, Ben-Hur H, Naor D, Nagler A, Lapidot T (2004) Blood 103:2981
35. Tabata T, Takei Y (2004) Development 131:703
36. Chen CS, Mrksich M, Huang S, Whitesides GM, Ingber DE (1997) Science 276:1425
37. Singhvi R, Kumar A, Lopez GP, Stephanopoulos GN, Wang DI, Whitesides GM, Ingber DE (1994) Science 264:696
38. Dike LE, Chen CS, Mrksich M, Tien J, Whitesides GM, Ingber DE. In Vitro Cell Dev Biol Anim 35:441
39. Chen CS, Ostuni E, Whitesides GM, Ingber DE (2000) Methods Mol Biol 139:209
40. Whitesides GM, Ostuni E, Takayama S, Jiang X, Ingber DE (2001) Annu Rev Biomed Eng 3:335

41. Ingber DE (2003) J Cell Sci 116:1157
42. Ingber DE (2003) J Cell Sci 116:1397
43. Cavalcanti-Adam EA, Tomakidi P, Bezler M, Spatz JP (2005) Prog Orthod 6:232
44. Arnold M, Cavalcanti-Adam EA, Glass R, Blummel J, Eck W, Kantlehner M, Kessler H, Spatz JP (2004) ChemPhysChem 5:383
45. Cukierman E, Pankov R, Yamada KM (2002) Curr Opin Cell Biol 14:633
46. Cukierman E, Pankov R, Stevens DR, Yamada KM (2001) Science 294:1708
47. Discher DE, Janmey P, Wang YL (2005) Science 310:1139
48. Engler AJ, Griffin MA, Sen S, Bonnemann CG, Sweeney HL, Discher DE (2004) J Cell Biol 166:877
49. Davis GE, Senger DR (2005) Circ Res 97:1093
50. Hinck L (2004) Dev Cell 7:783
51. Kanwar YS, Wada J, Lin S, Danesh FR, Chugh SS, Yang Q, Banerjee T, Lomasney JW (2004) Am J Physiol Renal Physiol 286:F202
52. Ramirez F, Sakai LY, Dietz HC, Rifkin DB (2004) Physiol Genomics 19:151
53. Birk DE, Bruckner P (2005) Collagen Suprastructures. In: Brinckmann J, Notbohm H, Müller PK (eds) Collagen: primer in structure, proscessing and assembly. Springer, Berlin Heidelberg New York, p 185
54. Adachi E, Hopkinson I, Hayashi T (1997) Int Rev Cytol 173:73
55. Engel J, Bächinger HJ (2005) Top Curr Chem 247:7
56. Canty EG, Kadler KE (2005) J Cell Sci 118:1341
57. Kadler KE, Hojima Y, Prockop DJ (1987) J Biol Chem 260:15696
58. Kuznetsova N, Chi SL, Leikin S (1998) Biochemistry 37:11888
59. Gelman RA, Poppke DC, Piez KA (1979) J Biol Chem 254:11741
60. Silver FH (1981) J Biol Chem 256:4973
61. Gelman RA, Williams BR, Piez KA (1979) J Biol Chem 254:180
62. Zaleskas JM, Kinner B, Freyman TM, Yannas IV, Gibson LJ, Spector M (2004) Biomaterials 35:1299
63. Yannas IV, Burke JF, Gordon PL, Huang C, Rubenstein RH (1980) J Biomed Mater Res 14:107
64. McPherson J, Sawamura SJ, Condell RA, Rhee W, Wallace DG (1988) Collagen Relat Res 8:65
65. Salchert K, Streller U, Pompe T, Herold N, Grimmer M, Werner C (2004) Biomacromolecules 5:1340
66. Wallace DG, Rosenblatt J (2003) Adv Drug Deliv Rev 55:1631
67. Giraud Guille MM, Mosser G, Helary C, Eglin D (2005) Micron 36:602
68. Sheu MT, Huang JC, Yeh GC, Ho H (2001) Biomaterials 22:1713
69. McCloskey KE, Gilroy ME, Nerem RM (2005) Tissue Eng 11:497
70. Timpl R, Brown JC (1996) Bioessays 18:123
71. Kühn K (1994) Matrix Biol 14:439
72. Timpl R, Wiedemann H, van Delden V, Furthmayr H, Kühn K (1981) Eur J Biochem 120:203
73. Yurchenco PD, O'Rear JJ (1994) Curr Opin Cell Biol 6:674
74. Kleinman HK, McGarvey ML, Hassell JR, Martin GR (1983) Biochemistry 22:4969
75. Adachi E, Takeda Y, Nakazato K, Muraoka M, Iwata M, Sasaki T, Imamura Y, Hopkinson I, Hayashi T (1997) J Electr Microsc 46:233
76. Montanez E, Casaroli-Marano RP, Vilaro S, Pagan R (2002) Angiogenesis 5:167
77. Kleinman HK, Martin GR (2005) Semin Cancer Biol 15:378
78. Kofidis T, de Bruin JL, Hoyt G, Lebl DR, Tanaka M, Yamane T, Chang CP, Robbins RC (2004) J Thorac Cardiovasc Surg 128:571

79. Mao Y, Schwarzbauer JE (2005) Matrix Biol 24:389
80. Phillips JB, King VR, Ward Z, Porter RA, Priestley JV, Brown RA (2004) Biomaterials 25:2769
81. Ahmed Z, Briden A, Hall S, Brown RA (2004) Biomaterials 25:803
82. Carmeliet P (2004) J Intern Med 255:538
83. Clark RA (1993) Am J Med Sci 306:42
84. Hsieh P, Chen LB (1983) J Cell Biol 96:1208
85. Lachlan J, Currie J, Sharpe L, Martin R (2001) Plast Reconstr Surg 108:1713
86. Kowalewski T, Holtzman DM (1999) Proc Natl Acad Sci USA 96:3688
87. Yang G, Woodhouse KA, Yip CM (2002) J Am Chem Soc 124:10648
88. Bergkvist M, Carlsson J, Oscarsson S (2003) J Biomed Mater Res 64A:349
89. Keselowsky BG, Collard DM, Garcia AJ (2003) J Biomed Mater Res 66A:247
90. Sherratt MJ, Bax DV, Chaudhry SS, Hodson N, Lu JR, Saravanapavan P, Kielty CM (2005) Biomaterials 26:7192
91. Jiang F, Khairy K, Poole K, Howard J, Muller DJ (2004) Microsc Res Tech 64:435
92. Haynes CA, Norde W (1995) J Colloid Interface Sci 169:313
93. Norde W, Lyklema J (1991) J Biomater Sci Polym Ed 2:183
94. Pompe T, Zschoche S, Herold N, Salchert K, Gouzy MF, Sperling C, Werner C (2003) Biomacromolecules 4:1072
95. Renner L, Pompe T, Salchert K, Werner C (2004) Langmuir 20:2928
96. Renner L, Pompe T, Salchert K, Werner C (2005) Langmuir 21:4571
97. Pompe T, Kobe F, Salchert K, Jørgensen B, Oswald J, Werner C (2003) J Biomed Mater Res 67A:647
98. Pompe T, Keller K, Mitdank C, Werner C (2005) Eur Biophys J 34:1049
99. Geiger B, Bershadsky A, Pankov R, Yamada KM (2001) Nat Rev Mol Cell Biol 2:793
100. Balaban NQ, Schwarz US, Riveline D, Goichberg P, Tzur G, Sabanay I, Mahalu D, Safran S, Bershadsky A, Addadi L, Geiger B (2001) Nat Cell Biol 3:466
101. Pankov R, Cukierman E, Katz Z, Matsumoto K, Lin DC, Lin S, Hahn C, Yamada KM (2000) J Cell Biol 148:1075
102. Pompe T, Renner L, Werner C (2005) Biophys J 88:527
103. Pompe T, Markowski M, Werner C (2004) Tiss Eng 10:841
104. Ingber DE (2002) Circ Res 91:877
105. Wissink MBJ, Beernink R, Pieper JS, Poot AA, Engbers GHM, Beugeling T, van Aken WG, Feijen J (2001) Biomaterials 22:151
106. Kuberka M, Heschel I, Glasmacher B, Rau G (2002) Biomed Technol 47:485
107. Di Lullo GA, Sweeny SM, Körkkö J, Ala-Kokko L, San Antonio JD (2002) J Biol Chem 277:4223
108. Salchert K, Pompe T, Sperling C, Werner C (2003) J Chrom A 1005:113
109. San Antonio JD, Lander AD, Karnovsky NJ, Slayter HS (1994) J Cell Biol 125:1179
110. Salchert K, Oswald J, Grimmer M, Herold N, Werner C (2005) J Mat Sci Mat Med 16:581
111. Netelenbos T, van den Borg J, Kessler FL, Zweegman S, Huijgens PC, Dräger AM (2003) J Leukoc Biol 74:1035
112. Deguchi T, Komada Y (2000) Leuk Lymphoma 40:25
113. Askenasy N, Zorina T, Farkas DL, Shalit I (2002) Stem Cells 20:301
114. Hubbell JA (2003) Curr Op Biotechnol 14:551
115. Rizzi SC, Hubbell JA (2005) 6:1226
116. Raeber GP, Lutolf MP, Hubbell JA (2005) Biophys J 89:1374
117. Lutolf MP, Weber FE, Schmoeker HG, Schense JC, Kohler T, Müller R, Hubbell JA (2003) Nat Biotechnol 21:513

118. Massia SP, Hubbell JA (1991) J Cell Biol 114:1089
119. Irvine DJ, Hue KA, Mayes AM, Griffith LG (2002) Biophys J 82:120
120. Brandley BK, Schnaar RL (1989) Dev Biol 135:74
121. Koo LY, Irvine DJ, Mayes AM, Lauffenburger DA, Griffith LG (2002) J Cell Sci 115:1423
122. Richardson TP, Peters MC, Ennett AB, Mooney DJ (2001) Nat Biotechnol 19:1029
123. Ehrbar M, Djonov VG, Schnell C, Tschanz SA, Martiny-Baron G, Schenk U, Wood J, Burri PH, Hubbell JA, Zisch AH (2004) Circ Res 94:1124

Adv Polym Sci (2006) 203: 95–144
DOI 10.1007/12_072
© Springer-Verlag Berlin Heidelberg 2006
Published online: 10 January 2006

Hydrogels for Musculoskeletal Tissue Engineering

Shyni Varghese (✉) · Jennifer H. Elisseeff

Department of Biomedical Engineering, Johns Hopkins University,
3400 N Charles Street, Baltimore, Maryland 21218, USA
shyni@jhu.edu, jhe@bme.jhu.edu

Abstract The advancements in scaffold-supported cell therapy for musculoskeletal tissue engineering have been truly dramatic in the last couple of decades. This article briefly reviews the role of natural and synthetic hydrogels in the above field. The most appealing feature of hydrogels as scaffolding materials is their structural similarity to extracellular matrix (ECM) and their easy processability under mild conditions. The primary developments in this field comprise formulation of biomimetic hydrogels incorporating specific biochemical and biophysical cues so as to mimic the natural ECM, design strategies for cell-mediated degradation of scaffolds, techniques for achieving in situ gelation which allow minimally invasive administration of cell-laden hydrogels into the defect site, scaffold-mediated differentiation of adult and embryonic stem cells, and finally, the integration of tissue-engineered "biological implants" with the native tissue. All these developments in regenerative medicine are reviewed in this article.

Keywords Hydrogels · Minimally invasive · Musculoskeletal tissue engineering · Scaffold · Stem cell

Abbreviations

ACT	Autologous Chondrocyte Transplantation
BMP	bone morphogenetic proteins
CO_2	carbon dioxide
CS	chondroitin sulfate
ECM	extracellular matrix
e-PTFE	expanded poly(tetrafluroethylene) membrane
ES	embryonic stem cells
GAG	glycosaminoglycan
HA	hyaluroic acid
hES	human embryonic stem cells
IVD	intervertebral disc
LCST	lower critical solution temperature
MIS	minimally invasive surgeries
MMP	matrix metalloproteinase
MSC	mesenchymal stem cells
NiTi	nitinol
OA	osteoarthritis
OPF	poly(ethylene glycol fumerate)
PAA	poly(acrylic acid)
PAMPS	poly(acrylamidomtheyl propane sulfonic acid)
PEG	poly(ethylene glycol)
PEO	poly(ethylene oxide)
PEGDA	poly(ethylene glycol diacrylate)
PGA	poly(glycolic acid)
PHEMA	poly(hydroxyl ethyl methacrylate)
PLA	poly(lactic acid)
PLGA	poly(lactic-glycolic) acid
PNIPAm	poly(N-isopropyl acrylamide)
PPF	poly(propylene fumerate)
PVA	poly(vinyl alcohol)
RGD	arginine-glycine-aspartic acids
SIS	small intestinal submucosa
2D	two dimensional
3D	three dimensional

1
Introduction

Regenerative medicine is an emerging, interdisciplinary field which integrates a number of diverse fields including cell biology, materials science, engineering, and clinical research. The primary purpose of regenerative medicine is to assist the body in regenerating diseased or damaged tissues and organs. Developments in regenerative medicine involving gene and protein therapy, advanced biomaterials, and tissue engineering have evolved to render new opportunities to treat and cure many diseases which were irremediable decades ago. Such therapies range from recombinant human proteins that effectively treat certain metabolic deficiencies to specialized ceramics and polymers that serve as orthopedic implants in various joint replacements and bone grafts. These developments in medicinal technology have extended the longevity of human beings by several decades over their past generations. Excellent reviews on these subjects may be found elsewhere [1-6]. The focus of this review is to evaluate the role of one particular subset of biomaterials in regenerative medicine, namely, the impact of hydrogels on musculoskeletal tissue engineering.

Recent years have witnessed an exponential growth in the field of tissue engineering as documented by Lysaght and Reyes [7]. The overarching principle of tissue engineering is to create "off-the-shelf" tissues and organs to potentially replace those compromised by trauma, injury, disease, or aging. Current therapies for damaged tissue restoration using artificial implants have several shortcomings such as suboptimal long-term outcomes, long-term presence of implants within the body, invasive surgical procedure and associated morbidity, risk of infection, structural failure, and implant rejection [8-11]. In addition, the materials from which these implants are constructed do not integrate well with the host tissue and have a limited lifetime [10]. Other strategies involving the use of allografts and autografts also suffer from various drawbacks such as lack of donor tissue availability, donor site morbidity, increased susceptibility to disease transmission and immunorejection [11-13]. Table 1 lists the number of surgical procedures that have been carried out in USA alone per year to rectify tissue and/or organ damage [14]. As of May 15, 2005, approximately 88 475 patients in USA are waiting for organ transplant whereas only 4367 patients have received transplants this year by April 29, 2005 [15]. This striking gap between donor supply and demand worsens year by year, and is predicted to widen even further with the aging of the baby-boomer generation. In fact, it is predicted that in USA alone, one person out of every five approaching the age of 65 will need some form of temporary or permanent organ replacement therapy at least once in their life time [16]. According to the Business Communications Company, Inc., the USA market for regenerative medicine products for bone and joint implants has been projected to approach 1.4 billion by 2007 [17]. However, there are still a number of challenges, both scientific and translational (manufacturing and regulatory), that must be overcome before widespread application of these technologies can be achieved.

The musculoskeletal system includes bone, cartilage, and spine which are exciting tissue engineering targets as the need for their replacement is

Table 1 Incidence of organ and tissue deficiencies, or the number of surgical procedures related to these deficiencies, in the United States. This is a partial list compiled from sources that include the American Diabetes Association, American Liver Foundation, Muscular Dystrophy Association, American Red Cross, American Kidney Foundation, The Wilkerson Group, Cowen and Co., American Academy of Orthopedic Surgery, American Heart Association, National Institute of Neurological Disorders and Stroke, Source Book of Health Insurance (Health Assurance Association of America) 1991, Federal Register, and Department of Health and Human Service (Medicare-based information). [Reprinted with permission from Langer R, Vacanti JP (1993) Science 260:920, copyright 1993 AAAS]

Indication	Procedures or patients per year
Skin	
Burns *	2 150 000
Pressure sores	1 500 000
Venous stasis ulcers	500 000
Diabetic ulcers	600 000
Neuromuscular disorders	200 000
Spinal cord and nerves	40 000
Bone	
Joint replacement	558 200
Bone graft	275 000
Internal fixation	480 000
Facial reconstruction	30 000
Cartilage	
Patella resurfacing	216 000
Chondomalacia patellae	103 400
Meniscal repair	250 000
Arthritis (knee)	149 900
Arthritis (hip)	219 300
Fingers and small joints	179 000
Osteochondritis dissecans	14 500
Tendon repair	33 000
Ligament repair	90 000
Blood vessels	
Heart	754 000
Large and small vessels	606 000
Liver	
Metabolic disorders	5000
Liver cirrhosis	175 000
Liver cancer	25 000
Pancreas (diabetes)	728 000
Intestine	100 000
Kidney	600 000

* Approximately 150 000 of these individuals are hospitalized and 10 000 die annually

Table 1 (continued)

Indication	Procedures or patients per year
Bladder	57 200
Ureter	30 000
Urethra	51 900
Hernia	290 000
Breast	261 000
Blood transfusions	18 000 000
Dental	10 000 000

high. Bone is one of the most common transplanted tissues, second only to blood [18]. Hence, the need for new bone engineering strategies is quite apparent although the tissue has a strong natural capacity for repair. Cartilage has also been the focus of significant tissue engineering research for the last decade given its lack of a self-repair ability when it is lost due to trauma, disease, or congenital abnormalities. Cartilage is an avascular tissue, which not only provides structural function as in the nose and ears, but the tissue also acts as a cushion and lubricating surface for proper joint function. Cartilage defects are not uncommon as over 1 million surgical procedures are performed annually to correct defective cartilage, with a significant fraction devoted to facial and dental restoration [19]. Like joint degeneration, degenerative spinal disorders are also very common, with 5.7 million people being diagnosed with intervertebral disc (IVD) degenerations every year in the United States [20]. The IVD consists of a gelatinous nucleus pulposus encased within a highly elastic annulus fibrosus attached to the bony vertebral tissue. The disc is avascular, receives nutrients by passive diffusion, and lacks the ability to self repair. The degeneration of nucleus pulposus, and subsequent disc herniation, often results in severe nerve compression symptoms ranging from lower back pain to temporary paralysis [20, 21]. Indeed low back pain is identified as one of the most common causes of musculoskeletal impairment [21]. These are just few examples which show the socio-economic necessity for tissue-engineered implants.

2
General Concepts in Tissue Engineering

Tissue engineering provides a solution to the earlier discussed inadequacies of nonbiological replacements as well as allografts and autografts. Tissue engineering approaches mainly consist of the following key components: cells, biomaterial scaffolds and growth factors or other biological signals. Musculoskeletal tissues that are currently being engineered using these key components include cartilage [22–27], bone [18, 28–32], muscle [33], tendon [34], and IVD [35–37]. The various tissue engineering approaches can be broadly classified into three main categories: (i) guided tissue regeneration utilizing the natural regenerative ability of the tissues; (ii) cell therapy using either allograft or autograft cells; and (iii) scaffold-supported cell therapy. Notably, polymers play a pivotal role in each of these approaches as scaffolding material which will be discussed below in more detail.

2.1
Biomaterials for Guided Tissue Repair

Guided tissue repair relies on the natural ability of the body to regenerate the lost tissue. In this approach a suitable polymeric scaffold is implanted into the defect site. The primary purpose of this scaffold is to provide mechanical support to enhance migration of cells, which have the ability to proliferate and produce the tissue matrix, from the neighboring tissues into the defect site as illustrated in Fig. 1. The biomaterial also acts as a barrier and occludes the entrance of undesirable cells. In more advanced approaches, biomaterials are also used to deliver growth factors and gene delivery vectors to the defect site so as to accelerate the regeneration of tissue. The growth factors diffuse into the neighboring tissue and bind to the cell receptors and accelerate their migration into the defect site. An ideal biomaterial for this purpose is expected to navigate the tissue regeneration process by providing necessary cues for

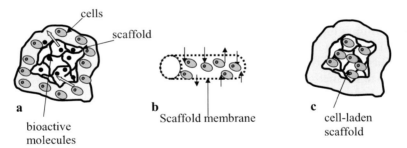

Fig. 1 a Guided regeneration of damaged tissue. A composite scaffold, containing hydrogels and bioactive molecules, is placed onto the defect site. The scaffold enhances the migration of progenitor cells from the surrounding tissue into the defect site and subsequently regenerates the lost tissue. The bioactive molecules diffuse out of the scaffolds and bind to the neighboring cells so as to accelerate the migration of cells into the defect site. **b** Cell therapy. Allograft cells are encapsulated within a closed polymer membrane system to allow exchange of nutrients, oxygen and biotherapeutic substances between the encapsulated cells and the surroundings. **c** Cell-scaffold system. A cell population which has the ability to produce the lost tissue is placed onto the defect site after encapsulating within a promising scaffold system

a desired cell function, and degrade as the tissue repair proceeds. Regeneration of skin using a porous hydrogel scaffold containing glycosaminoglycan (GAG) and collagen is an example where the implanted scaffold stimulates the self repair ability of our bodies to regenerate the lost tissue [38–40], and this material received FDA approval in 1996 [41]. This product commercially known as *Integra* (by Integra life science and Johnson & Johnson) is used for skin burn treatments. In another study, Ono et al. have used a combination of polymers and growth factors to accelerate the healing process in bio-interactive wound dressings [42].

Guided tissue engineering with the aid of acellular biomaterials has also been explored for bone regeneration. One of the polymeric materials that has been widely used to promote osteogenesis and periodontal regeneration is expanded poly(tetrafluroethylene) membrane (e-PTFE or Gore-Tex) which has a useful microporous structure and is biocompatible [43, 44]. The drawback of this material is that it is bio-inert and nondegradable. Bio-degradable polymers that have been used for guided regeneration of bone include polylactide, polyglycolide, and their copolymers, biomimetic peptide hydrogels, collagen-based matrices, and poly(propylene fumerate) [44–48]. Acellular porcine small intestinal submucosa (SIS) has been extensively studied as a xenogenic scaffold for tissue repair and it is a bio-resorbable material [49]. This material recently gained FDA approval for skin repair, soft tissue support, and rotator cuff repair [49, 50]. The success of SIS as a scaffold navigator for regeneration is attributed to the presence of ECM proteins, and other growth factors such as fibroblast growth factor (FGF-2) and transforming growth factor β (TGF-β) [51].

2.2
Cell Therapies

Cell therapy, serendipitously discovered by Paul Niehans in 1931, involves the transplantation of allograft and autograft cells to repair dysfunctional tissues. Since then, procedures attempting to utilize allograft as well as autograft cells that are capable of producing biotherapeutic substances have been exploited to treat diabetes, liver failure, and neural disorders such as Alzheimer's diseases, Parkinson's disease, spinal cord injuries, etc. [52–55]. However, such an approach needs to be reconsidered for transplantation of allograft and xenograft cells owing to their vulnerability to immunorejection. In order to circumvent immunorejection, allograft and xenograft cells are often encapsulated within a permeable polymer membrane. The main role of the polymer membrane is to prevent the entrance of high molecular weight immuno-responsive agents into the implant site. The encapsulated cells are then expected to produce the required biotherapeutic substance which will treat the disease (see Fig. 1). The permeable membrane also allows the exchange of nutrients, oxygen and biotherapeutic substances between the encapsulated cells and the surroundings [54]. The permeability, configuration, and chemistry of the membranes have been tailored successfully by various investigators to create membranes with optimal properties. Two commonly investigated polymeric systems include hollow fiber membranes of poly(acrylonitrile)-poly(vinyl chloride) and microcapsules of alginate [5, 54].

Other than treating the above-mentioned diseases, cell therapy has also been investigated for treating tissue loss in the musculoskeletal system. For example, Autologous Chondrocyte Transplantation (ACT), Carticel (Genzyme Biosurgery), an FDA approved product, utilizes cells alone to regenerate lost articular cartilage. Here, the cells from the patients are first isolated from a biopsy and then expanded ex vivo prior to their re-implantation [56]. Such an approach involving ACT to repair musculoskeletal lesions (e.g. large chondral defects of the knee) was first introduced in Sweden in 1987. Since then various researchers have investigated the success of ACT, and found that the average success rate of repairing various lesions (such as femoral chondyle, osteochondritis dissecans, etc.) using this technique is approximately 85% [57–59].

2.3
Bioscaffolds and Cells

When the defect sites are large and the cell migration from the residual surrounding tissues is minimal or impeded, an implantation of both cells and a functional scaffold is necessary for repair. In this approach, the scaffold provides the initial structural support to the encapsulated cells, and guides their proliferation and differentiation into the desired tissue or organ. A cell population which has the ability to proliferate and produce the required matrix is placed onto the defect site in combination with a biomaterial as shown in Fig. 1. A number of cell sources exist for this purpose, and these include fully differentiated cells isolated from tissue (e.g. chondrocytes for cartilage repair), adult stem cells (e.g. bone marrow-derived mesenchymal stem cells, MSCs) which are multipotent, or embryonic stem cells (ES) which are pluripotent. The polymer scaffold acts as an artificial extra-cellular matrix (ECM) and provides a favorable niche or microenvironment for the cells to grow towards the desired lineage. Additionally, it also serves as a carrier for the transport of cells into the defect site and also confines the cells to the defect site. Cells can be either seeded onto a solid fibrous or porous scaffold or encapsulated within a gelatinous scaffold. In both cases, the cells are suspended or attached to the scaffold, and then they proliferate, migrate, and secrete ECM.

The chemistry as well as the biomechanical properties of the scaffold plays a key role in the differentiation of stem cells to the particular lineage such as cartilage, bone, or even muscle. Differentiation of adult and embryonic stem cells is generally controlled by various cues from the microenvironment, which will be discussed in more detail later. In short, designing a proper scaffold is a pivotal issue in tissue engineering, and much progress has been made in this pursuit over recent years, with the development of smart biomimetic scaffolds being the key contenders [60]. Examples of tissue engineered products using this strategy include Apligraf, Dermagraft, and Orcel, all of which consist of dermis cells and collagen. These products are mainly used for skin replacement, in the case of burns and diabetic ulcer. This approach has also been used for regenerating other tissues such as cartilage, e.g. Hyalograft C which uses a combination of chondrocytes and a hyaluronan (HA)-based scaffold [61].

3
Scaffolds in Tissue Engineering

As discussed above, in order to create biologically and functionally active tissue/organ replacements, cultured cells are grown on or within a scaffold. The scaffold is expected to provide a specific biological and mechanical environment to the encapsulated cells. The scaffold also assigns a predefined architecture to the regenerated tissue. Furthermore, various growth factors and other bioactive signals and bio-molecules can be loaded into the scaffold along with the cells to guide the regulation of cellular functions during tissue development [62–65]. Scaffolding biomaterials can be designed to meet a series of stringent requirements as described below that are either required or highly desirable to optimize tissue formation.

1. The matrix should be biocompatible and should promote cell growth.
2. Scaffolds that are designed to encapsulate cells must be able to solidify without damaging the cells.
3. The scaffold must allow diffusion of nutrients and metabolites between the encapsulated cells and the surroundings.
4. The scaffold material should degrade in response to the production of ECM components into noncytotoxic segments for easy elimination. The degradation of the scaffold also prevents the scaffolds from impeding the growth of new tissue.

Tissue engineering scaffolds are constructed out of natural and synthetic polymers. Natural polymers typically comprise of collagen or other proteins, polysaccharides, fibrin, or other biomolecules. The use of these polymers to construct hydrogels is discussed later in Sect. 4.1. Synthetic biodegradable polymers used for tissue engineering scaffolds involve FDA-approved polyesters such as poly(lactic acid) (PLA), poly(glycolic acid) (PGA), and their copolymer poly(lactic-glycolic) acid (PLGA) (refer to Fig. 2 for their chemical structures) [66].

Vacanti et al. have reported that poly(glycolic acid) matrices are capable of supporting the regeneration of musculoskeletal tissues such as cartilage

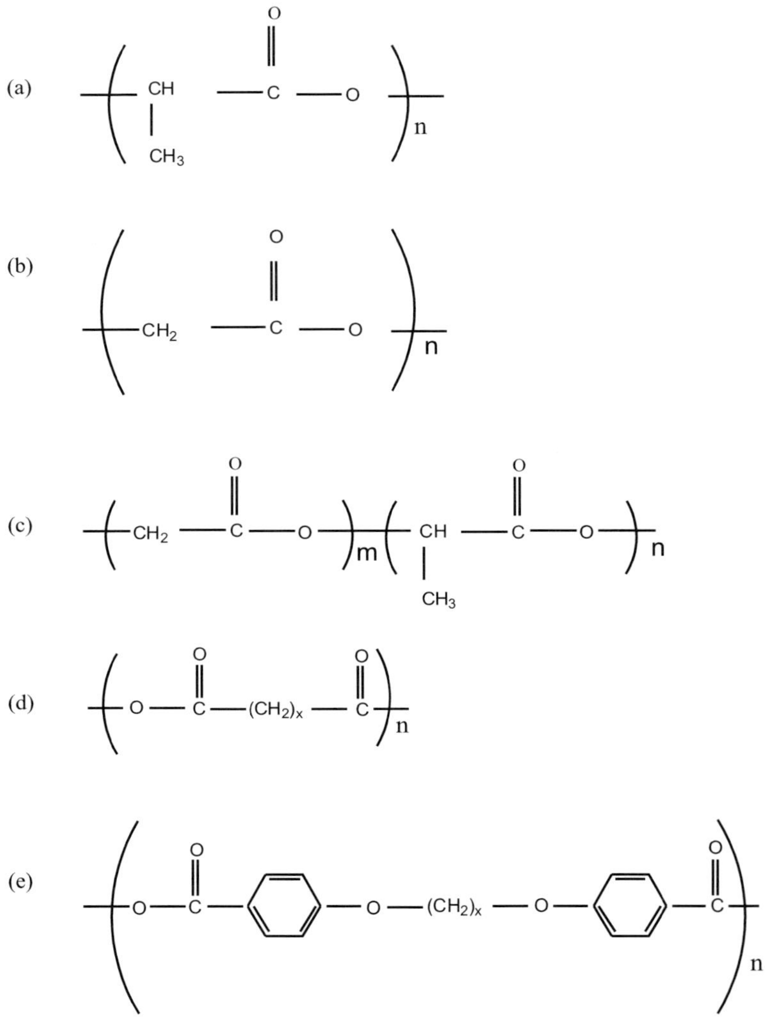

Fig. 2 Chemical structure of **a** poly(lactic acid), **b** poly(glycolic acid), **c** poly(lactic-co-glycolic) acid, **d** poly(anhydride), and **e** poly(imide)

and bone [66, 67]. In addition to scaffolding, they have also been widely used for controlled release of drugs [62, 63, 65, 68, 69]. A combination of the above polymers along with growth factors has successfully been used for guided bone regeneration in vivo by various investigators [70, 71]. These polymers possess good mechanical properties and the ester linkages present on the polymer backbone are labile to hydrolytic degradation. Because of their excellent mechanical properties, biocompatibility and degradability these polymers, belonging to the poly(α-hydroxy acid) family, have also been exploited to control the differentiation of human ES (hES) cells into the musculoskeletal lineage [72]. However, these advantages are overshadowed by the fact that their acidic degradation products provoke an immunoresponse from the body [73].

Another class of synthetic biomaterial that has been extensively explored for biomedical applications such as controlled release of drugs include poly(anhydrides) [74, 75]. These materials are useful for drug delivery because of their inherent surface erosion mediated degradation, unlike their polyester counterparts which undergo bulk degradation [76]. The anyhydride bonds present in the polymer backbone can be hydrolyzed by the aqueous milieu of the body when it is implanted in vivo. Poly(anhydride) has also been approved by the FDA (Gliadel) for the delivery of carmustine for treating brain cancer [76, 77]. Poly-Aspirin is another novel poly(anhydride) which can be degraded into salicylic acid in order to reduce inflammation and pain locally [78].

The strong mechanical properties and degradation kinetics of poly(anhydrides) have attracted many researchers to investigate their potential as a bone tissue engineering scaffold. According to Ibim et al. poly(anhydride-co-imide) polymers are comparable to PLGA in terms of their biocompatibility and they are also capable of supporting critical bone regeneration [79]. Recently, Anseth et al. have developed photocrosslinkable poly(anhydrides) for bone regeneration where polymerization can be achieved in situ, which makes them suitable for minimally invasive applications [80]. Although poly(anhydrides) offer some advantages, their applicability in delivering cells to the defect site is limited because of their highly crosslinked (dense) polymer network structure which reduces cell viability [81].

Scaffold matrices may possess a "quasi" two dimensional (2D) geometry, where the cells are seeded *onto* the scaffolds, or a three dimensional (3D) geometry, where the cells are seeded *within* the scaffold. Recent findings have demonstrated that porous three-dimensional scaffolds are superior to their two-dimensional counterparts in providing a proper physiological environment to the cells [82, 83]. When cells are seeded within three-dimensional scaffolds, the ECM proteins produced by the cells are deposited uniformly within the scaffold, which then remodel to yield the desired cytoarchitecture.

Porous 3D polymeric networks with interconnected pores to allow cell growth, vascularization and diffusion of nutrients are created by various methods. Freeze drying has been utilized to create porous collagen-GAG composite matrices for skin regeneration by Chen et al. [84], whereas Sastry et al. have used hydrocarbon templating to create porous scaffolds [85]. A gas-foaming process using carbon dioxide (CO_2) as the foaming agent has also been used to fabricate highly porous polymer structures [86–88]. Salt leaching is another approach for producing structures similar to foam or sponge where the size of the pore is controlled by the size of the salt crystals [89]. Both woven and unwoven fibers have also been commonly used to construct a porous scaffold structure [90]. In this case, the fiber diameter and the distance between the fibers are adjusted to meet the requirements of the scaffold porosity. The structural and mechanical properties of the scaffolds are significantly influenced by their pore size. These parameters also greatly influence the functionality of the engineered tissue by influencing cell growth, differentiation, and tissue organization [18, 91]. Ma et al. have used thermally induced phase separation followed by subsequent sublimation to create highly porous three-dimensional scaffolds for bone tissue engineering [18]. In this approach, the morphology and size of pores can be varied by changing the solvent, polymer concentration, and phase separation temperature. This method has also been used to create macroporous scaffolds with pore diameter which are amenable to osteoblastic cells and bone tissue growth [18].

Another important criteria in using any biomaterials as a tissue engineering scaffold for organ replacement is that the scaffold should be able to assume the anatomical shape and structure of the targeted tissue or organ. For example, tube-shaped scaffolds are used to engineer tubular tissues like arteries [92]. Cao et al. have engineered a three-dimensional elastic cartilage graft by seeding chondrocytes onto a prefabricated (ear shape) PLGA scaffold, where the scaffold directed the ultimate shape of the neocartilage tissue [93]. Shastri et al. have used a different approach based on a hydrocarbon-templating process to create three-dimensional PLA scaffolds with specific shapes to regenerate tissues with pre-defined shapes [85].

Most of the solid scaffolds discussed above are hydrophobic and are processed under severe conditions, which make further modifications such as incorporation of biochemical and biophysical functions a serious challenge. They also need to be prefabricated and are therefore not ideal for irregularly shaped defects. Furthermore, these scaffolds need to be surgically implanted through invasive surgical procedures [93]. To this end, viscoelastic hydrogels have been employed as alternative scaffolding materials because of the many advantages they offer over hydrophobic polymer networks as discussed in the following sections.

4
Hydrogel Scaffolds

Hydrogels are three-dimensional networks of hydrophilic polymers which have the ability to imbibe a large quantity of water and biological fluids. The network is formed through either chemical crosslinking (covalent and ionic) or physical crosslinking (entanglements, crystallites, and hydrogen bonds) as shown in Fig. 3. The elastic network holds the solvent inside the matrix by osmotic forces, while the liquid prevents the polymer network from collapsing into a compact mass. The combination of these two parameters, namely,

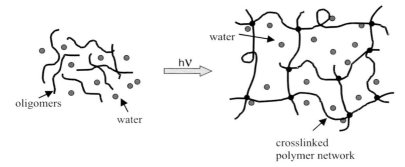

Fig. 3 Schematic of hydrogel formation. Crosslinked network is created through physical associations, ionic bonds, or covalent bonds

the osmotic forces and the elastic retractivity, defines the properties of the gels. Depending upon their chemical composition, crosslink density, and hydrophobicity, hydrogels can vary in consistency from viscous fluids to fairly rigid solids. Basically, hydrogels are wet, soft, and capable of undergoing large deformation. Thus, hydrogels are very similar in nature to mammalian tissues, which are essentially aqueous gels made up of proteins and polysaccharide networks. The hydrogel-like nature of tissues helps the organisms to transport oxygen, nutrients, and other bioactive moieties easily and effectively while retaining their solid nature.

Chemically crosslinked hydrogels are usually created by radical polymerization which enjoys the presence of a vinyl moiety in the starting monomer and oligomer mixtures [94, 95]. Hydrogels have also been created by taking advantage of the reaction between different functional groups present in the reacting oligomers [96–98]. Recently, Sperinde et al. have reported the formulation of a novel PEG-based hydrogel in which the network is formed by enzymatic crosslinking which uses transglutaminase, a natural enzyme, as a catalyst [99, 100]. Tranglutaminase is a calcium-dependent enzyme and it is ubiquitous in body fluid [99]. Another approach that has been widely used to create chemically crosslinked hydrogels involves photopolymerization through exposure to UV light [101], visible light (Varghese and Hill, 2005, personal communication), and γ-radiation [102]. Other approaches involving ionic interactions [103, 104], hydrophilic-hydrophobic interactions [105–108], crystallization [109–111], hydrogen bonding [112, 113], molecular recognition [114, 115], and self-assembly [116–120] have been widely explored to create physically crosslinked hydrogels.

Conventional hydrogels are known for their super-absorbing properties and they have found applications in personal care products and agricultural aids [121, 122]. There is another class of hydrogels known as "smart" hydrogels. The smartness of these hydrogels is characterized by their ability to respond to external stimuli such as temperature, pH, chemical residues and solvent [105, 121]. The characteristic features of these hydrogels have been creatively utilized for various biomedical and pharmaceutical applications [123, 124]. For example, hydrogels have been tailored to produce smart sensors that can detect small levels of blood-glucose [125], and as in situ insulin pumps to control glucose levels [126]. In addition, smart polymeric gels have also been considered as model biomimetic materials since they demonstrate several attributes of biomolecules such as sensitivity, selectivity, mobility, shape memory, self-organization, and healing [115, 127–130].

Recent years have witnessed a surge of interest in using bio-mimicking hydrogels as tissue engineering scaffolds. This increased interest in hydrogels for tissue engineering applications is attributed to their water content, viscoelasticity akin to the native tissues, biocompatibility, and their ability to permit diffusion of nutrients and bioactive molecules. In addition to the structural stability, the mechanical and biological properties of hydrogels can also be easily molecularly tailored to incorporate suitable biological signals that can stimulate cell proliferation and differentiation. The porosity of hydrogel can be easily tailored via controlling the crosslink density, monomer concentration, and molecular weight of the precursor. Hydrogels are considered as good candidates for in vivo applications because of their hydrophilic nature which resists protein and cell adhesion to the implant surface. Moreover, the soft viscoelastic nature of hydrogels minimizes irritation to the surrounding tissues as well as prevents stress shielding [95, 131, 132]. Anchorage independent cells like chondrocytes exhibit good cell viability within hydrophilic scaffolds like hydrogels [133]. Moreover, many investigators have shown that the hydrophilicity of the scaffold facilitates the re-differentiation of de-differentiated human nasal chondrocytes [134].

Another attractive feature of hydrogels is that they can be delivered in a minimally invasive manner as mentioned earlier. This is a very important aspect in tissue engineering approaches (for clinical applications) which has received a lot of attention recently, since hydrogel-cell systems can be injected into the body in a noninvasive manner for arthroscopic surgery. In our laboratory, we use this approach extensively for musculoskeletal tissue engineering applications. A detailed discussion on the minimal invasive application of polymers in the biomedical field is provided later in Sect. 6. Hydrogels can be broadly divided into two categories—natural and synthetic hydrogels, based on their source.

4.1
Natural Hydrogels

Natural hydrogels are water-liking networks that are made out of naturally derived polymers, and are widely used in tissue engineering. For example, extra cellular matrix components (ECM) such as hyaluronic acid (HA), chondroitin sulfate (CS), matrigel, and collagen-based hydrogels are used as scaffolds for cartilage and skin regeneration. Other naturally derived materials used to synthesize hydrogels include alginate, chitosan, fibrin, agarose, and silk.

Alginate: Alginate is an anionic polysaccharide found in seaweed, which is composed of β-D-mannuronic and α-L-guluronic acid as shown in Fig. 4. An appealing feature of alginate is that it undergoes gelation under physio-

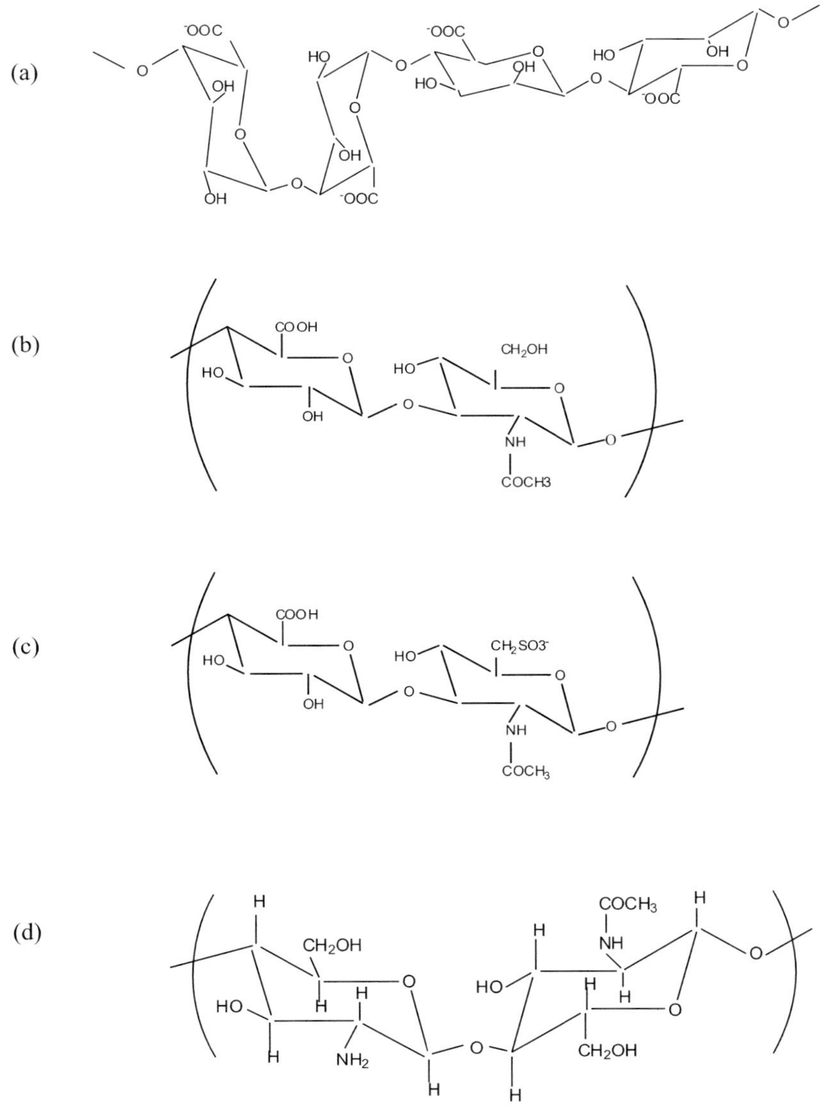

Fig. 4 Chemical structure of natural polymers that have been used to create hydrogels: **a** Alginate, **b** Hyaluronic acid, **c** Chondroitin sulfate, and **d** Chitosan

logical conditions, in the presence of a small concentration of certain divalent cations such as Ca^{2+}, Ba^{2+}, and Sr^{2+}, through the ionic interaction between the carboxylic group located on the polymer backbone and the cation. The metal-induced gelation of alginate is attributed to the ability of guluronic acid chains to form an "egg box"-shaped structure in the presence of alkaline earth metals [135]. This characteristic feature (cation-induced gelation) of alginates can be harnessed for minimally invasive applications. However, such ionically crosslinked alginate hydrogels mainly undergo degradation by uncontrolled dissolution of polymer chains. Some of these resulting chains have a high molecular weight and therefore cannot be readily eliminated by the body. In order to achieve bio-resorbable dissolution products, high molecular weight alginate chains are broken into smaller segments by treating them with γ-irradiation prior to their gelation [136]. Irradiated alginate hydrogels were found to support bone formation in vivo while allowing the ultimate excretion of the polymer chains by kidneys [30]. In another approach, partial oxidation of alginate polymer chains has been used to render them hydrolytically degradable [137]. In addition to metal induced gelation, alginate hydrogels can also be prepared by crosslinking alginate with other monomers [104, 138, 139].

Because of their biocompatibility and nontoxic nature, alginate beads have been used to deliver chondrocytes [140–142], hepatocytes [143], and islets of langerhans [54] into the body for cell therapy. Alginate has also been previously used for wound dressing [144] and as a scaffold for musculoskeletal tissue engineering [30, 145–148]. Chondrocytes are known to de-differentiate into fibroblast-like cells during their in vitro monolayer expansion, and alginate gels have been used to effectively re-differentiate the de-differentiated chondrocytes, where the alginate hydrogels provide a three-dimensional environment to the anchorage independent chondrocytes [140]. The chondrocytes that are encapsulated within the alginate hydrogels retain their spherical cell morphology and produced the cartilage-specific markers: collagen type II and aggrecan. Despite its advantages, the use of alginate as an ideal scaffold for tissue engineering is limited due to several shortcomings such as weak mechanical properties, lack of cellular interactions, and uncontrollable degradation profile. The incorporation of oligopeptides such as arginine-glycine-aspartic acids (commonly known as RGD peptides) has been utilized to enhance the adhesive property of alginate to some extent [145–147].

Collagen: Collagen is the main component of natural ECM and it is found in many tissues such as bone, tendon, skin, ligament, and other connective tissues. In particular, bone and teeth are made from a collagen-hydroxyapaptite composite. Collagen is composed of three protein chains wrapped around each other in a tight triple helix, which entangle with other helices via secondary interactions to form a thermally reversible hydrogel. The excellent biological properties of collagen have been used in various applications such as artificial skin (e.g. collagen is a component of FDA-approved wound dressing matrices), scaffolds for engineering liver, skin, bone, cartilage, and blood vessels [149–153], and as coatings for bio-inert implants [154]. For these applications, collagen is usually processed into various forms such as sponges, foams, fibers, and gels. The degradation of collagen matrix can be controlled locally by the cell-secreted metalloproteases such as collagenase [104].

Collagen hydrogels are excellent candidates for musculoskeletal tissue engineering scaffolds because cells can easily adhere onto the gels, and the scaffold can provide appropriate biological signals to the cells. Both collagen I and collagen II are often employed as a scaffold for cartilage engineering [155–159]. Collagen gels are known to provide a suitable milieu to chondrocytes so as to retain their phenotype (spherical morphology) in vitro [159]. In a detailed investigation, Nehrer et al. have compared the behavior of canine articular chondrocytes in collagen type I and type II sponges, and their studies showed that type II collagen sponges facilitate the chondrocyte phenotype, while they de-differentiated into fibroblastic cell morphology when seeded into collagen type I [155]. A similar observation regarding the morphology ("fibroblastic" morphology) was observed when adipose-derived stem cells were seeded into collagen type I scaffold and then cultured in chondrogenic medium [146]. Interestingly, irrespective of their morphology, the cells contained within collagen type I scaffold showed higher biosynthesis compared to scaffolds made out of agarose and alginate as quantified by collagen and GAG synthesis. A similar trend was observed by Qui et al. where the authors showed a higher proteoglycan synthesis by chondrocytes in collagen type II gels [160]. These results indicate the advantage of having a biologically active scaffold for tissue engineering.

Collagen hydrogels have also been used for bone regeneration. For example, Lindsey et al. have shown that the use of collagen gels as a space-filling agent can facilitate the repair of calvarial defects in a rat model [161]. A composite of Helistat, a crosslinked network of bovine type I collagen, and BMP-2 has been used for osseous regeneration in rabbits with unilateral critical-sized defects in the radii [162]. Helistat with BMP-2 has shown significant bone regeneration (comparable to the one treated with autografts) in 8 weeks time, whereas untreated defects or those treated with Helistat alone showed hardly any new bone formation.

Although the collagen-based hydrogels have numerous advantages, their use as tissue-engineering scaffolds is limited due to their inherent physical weakness. In order to enhance their mechanical properties, chemical crosslinks are often introduced using various crosslinkers [163] and enzymatic reactions [164]. Rault et al. have investigated the effect of various crosslinkers on the mechanical properties of collagen networks, and found that they depend strongly upon the crosslinker [165].

Hyaluronic Acid (HA): HA, also known as hyaluronan, is composed of *N*-acetyl-*D*-glucosamine and *D*-glucuronic acid as shown in Fig. 4. It is found in all mammalian tissues and body fluids, and is highly hygroscopic in nature. The excellent lubrication property of mammalian joints with minimal coefficient of friction is mainly attributed to the presence of HA in the synovial fluid within the joints. HA is also found in the human eye (aqueous humor) and the viscoelastic property of HA helps the proper functioning of the eye [166]. The viscoleastic property of HA has been widely exploited for ocular surgery and for alleviating osteoarthritis (OA) through its intraarticular administration [167]. In a recent study, Langer and coworkers have utilized this property of HA to create vocal cord replacements [168]. HA also serves as a free radical scavenger and as an antioxidant [169]. Their ability as a free radical scavenger has been hypothesized to enhance their role in treating symptoms of OA. HA is hydrophilic and its nonadhesive property make it the primary component of Seprafilm, a postoperative adhesive used in surgical applications. In addition to preventing protein adsorption, unmodified HA also restricts cells adhesion because of their smooth surface and anionic (hydrophilic) nature [170]. In order to increase cell adhesion, Ramamurthy et al. have applied UV treatment to induce topographical changes to HA, which subsequently enhances the attachment of neonatal rat smooth muscle cells onto the HA materials [170].

HA is prevalent during wound healing and plays a significant role during morphogenesis, embryonic development, and angiogenesis [171]. HA supposedly promotes the early inflammation (a crucial criterion for the initial wound healing) which then deteriorates slowly with time to allow matrix stabilization [171]. Not surprisingly, HA has been identified to have a pivotal role in tissue repair, and this property has been widely exploited for wound healing applications [171–173]. The potential of HA to be a unique biomaterial is limited due to the fact that native hyaluronan can be rapidly metabolized in vivo by free radicals and Hyaluronidase. Various methods such as covalent crosslinking and chemical modifications, which exploit the carboxylic and hydroxylic groups located on the HA backbone, have been adopted to increase the longevity and mechanical properties of HA while still maintaining its superior biological properties [166, 174–176]. For example, benzyl ester of Hyaluronan (commercially known as HYAFF 11 by Fidia Advanced Biopolymers) has been extensively explored as a tissue-engineering scaffold. The fact that nasoseptal de-differentiated chondrocytes seeded onto HYAFF reexpressed collagen type II both in vitro and in vivo, indicates that HA has a favorable effect on chondrocyte phenotype [177].

HA is also known to interact with chondrocytes via the surface receptor CD44, and this receptor-mediated signaling enables the chondrocytes to retain their phenotype [178]. The presence of HA in the culture medium has been used to enhance the proliferation of chondrocytes while keeping their phenotype intact [179]. HA-based materials are therefore extensively used in cartilage tissue engineering [180–183]. Hyalograft C is a tissue-engineered cartilage graft where the chondrocytes are seeded within a Hyaff 11 scaffold for cartilage regeneration [61]. Hyaff, has also been used as a scaffold for skin regeneration. Hyalomatrix PA, Hyalograft 3D, and Laserskin are just a few of the hyaluronan-based commercial skin grafts available for wound dressing.

Chondroitin Sulfate (CS): CS occupies around 80% of glycosaminoglycan (GAG), a major component of articular cartilage. CS is a disaccharide composed of glucuronic acid and *N*-acetylgalactosamine as shown in Fig. 4. An oral intake of CS has been identified to alleviate the symptoms of OA, and is commonly sold in USA as a dietary supplement. CS-based hydrogels have been used for cartilage tissue regeneration in order to achieve enhanced cell proliferation and proteoglycan secretion [184–186]. Sechriest et al. have shown that a monolayer of bovine articular chondrocytes plated on CS-chitosan composite expressed focal adhesions and maintained the characteristic chondrocyte phenotype [187]. Chang et al. have applied a tri-copolymer scaffold containing gelatin, CS and HA as a scaffold for cartilage tissue engineering [188]. In our laboratory, we have modified CS by incorporating methacrylate groups such that it can be photo-polymerized into a hydrogel and therefore can be used as a minimally invasive tissue engineering scaffold for cartilage replacement [185].

Similar to HA, CS has also been explored for wound healing applications in the past. Prestwich and coworkers have shown that wound dressings containing both HA and CS enhance the re-epithelization of the wound [173, 189]. Like HA and collagen, CS is also physically weak and degrades rapidly in vivo, but the formation of a hydrogel improves its mechanical properties and degradation rate. Other than being used as a scaffold component, our laboratory has also explored the possibility of CS-based polymers as a tissue adhesive for ophthalmology and for integrating the engineered cartilage to host tissue in the defect site (Wang and Elisseeff, 2005, personal communication) [98].

Fibrin: Fibrin is well known for its hemostatic function of preventing bleeding by forming blood clots, which are structurally similar to hydrogels. Hence fibrin has been considered as a natural scaffold for supporting wound healing. This property also makes fibrin glue a global surgical sealant, and it has been widely used as an adhesive in plastic and reconstructive surgery. Fibrin glue (hydrogel) is formed by the enzymatic polymerization of fibrinogen and thrombin at physiological conditions. The thrombin concentration present during the gelation influences the structural property of fibrin hydrogels and the extent of cell invasion [190, 191].

The most attractive feature of fibrin-based materials is their biological activity, and many studies have investigated the possibility of using them as scaffolds for various tissue engineering applications such as cartilage, bone, cardiovascular, and chronic wound healing [181, 192–194]. Various research groups have used fibrin glue for cartilage tissue engineering approaches [195–198]. Silverman et al. investigated the possibility of using fibrin glue for developing neocartilage in vivo in an athymic mouse and the biochemical analysis has confirmed that the ultimate tissue formed had cartilage-like properties [199]. Additionally, the adhesive property of fibrin has also been utilized for skin repair, bone grafts, and targeted immobilization of various cells [181, 200, 201]. Horch et al. utilized Fibrin glue (Tisseel) for keratinocyte transplantation in burn patients and their findings suggest the feasibility of using fibrin glue to treat chronic wounds [202]. Recently Curri et al. reviewed the role of fibrin glue in the development of tissue engineered skin grafts [203]. One of the major concerns of using fibrin glue is that it poses a potential immunological threat since it can carry life-threatening viruses such as HIV. In order to prevent this potential immunologic threat, fibrin can be produced from the patient's own blood to use as an autologous scaffold.

Chitosan: It is a linear polysaccharide composed of *D*-glucosamine and *N*-acetyl-*D*-glucosamine residues (see Fig. 4 for chemical structure) and is derived from chitin, a polysaccharide found in the exoskeleton of shellfish. Chitosan is semi-crystalline in nature and its crystallinity varies with the degree of deacetylation. Chitosan can also be used in a minimally invasive manner because it can undergo thermal and pH-triggered gelation [204], and is biodegradable as it can be enzymatically degraded in vivo by lysozyme and chytosanasitase enzymes. The biocompatibility and cationic nature of chitosan has been explored for a variety of biomedical applications such as gene delivery [205], wound dressings [206–208], and space-filling implants [209, 210].

Chitosan is structurally similar to GAG and hence has been widely used as a cartilage scaffold for tissue engineering applications in recent years [211–214]. According to Lu et al. intra-articular injection of chitosan increases the chondrocyte proliferation and maintains the cartilage thickness [215]. Chitosan-coated surfaces have been used for the expansion of human osteoblasts and chondrocytes [212]. Results from these studies have shown that a chitosan-coated surface promotes good cell viability and the surface enables the cells to retain their characteristic morphology similar to that observed in vivo. Additionally, osteoblasts expanded on chitosan film expressed significant collagen type I gene markers whereas chondrocytes expressed collagen type II and aggrecan. Matthew and coworkers have extensively studied the effect of CS-chitosan composite on chondrocyte phenotype and their observations substantiate the earlier described findings [213].

Hoemann et al. have investigated the potential of using injectable chitosanbased hydrogels for tissue engineering cartilage both in vivo and in vitro [216]. Buschmann and coworkers have developed a chitosan/blood-clot composite scaffold (commercially known as CarGel) for cartilage regeneration and has just finished pre-clinical trials [217]. In addition to their potential for chondrogenesis, chitosan hydrogels have also been investigated for bone regeneration [212, 218]. Muzzarelli et al. has successfully created mineralized bone-like tissues in osseous defects in rats, sheep, and dogs with the aid of chitosan-based scaffolds [218]. Chitosan has also been explored as a biological adhesive and a space-filling scaffold [219]. The space-filling property of chitosan is mainly attributed to its adhesive property [216, 219], ability to accelerate wound healing [217, 219], and excellent immunological activity [220]. For instance, photocrosslinked chitosan hydrogels have been used previously to seal pin holes in the small intestines, aorta, and trachea of mice [219].

Matrigel: The matrigel matrix (commercially available from BD Biosciences) is a soluble form of basement membrane extracted from mouse tumors that contain several components of ECM proteins. The major components of matrigel include laminin, collagen 1V, heparan sulfate proteoglycans, and entactin. At room temperature, matrigel polymerizes to produce biologically active hydrogel resembling the mammalian cellular basement membrane. Cells are known to behave as they do in vivo when they are cultured in matrigel matrix, and hence matrigel has been used as a model system to study cell behavior in a 3D environment [221]. Matrigel indeed provides a physiologically relevant environment for studying cell morphology, biochemical function, cell migration and invasion, and gene expression. Therefore, matrigel has been used to culture a wide variety of cells [72, 77]. In the case of rhesus monkey ES cells, the use of matrigel has been shown to induce cell growth and differentiation [222]. In another study, Xu et al. showed that the presence of matrigel along with mouse fibroblast conditioned medium maintains human ES cells in an undifferentiated state in a feeder-free culture system [223].

Silk: In contrast to other natural materials, silk scaffolds are mechanically robust, and have been used in the medical field as suture materials for centuries. Silk obtained from both silkworm (*Bombyx mori*) and dragline spider (*Nephilia Clavipes*) has been extensively studied for decades. Spider silk has drawn a lot of attention from various fields due to its exceptional mechanical and thermal properties. The modulus and tensile strength of spider silk fibers are very much comparable to some of the strongest man-made fibers such as Kevlar [224]. An interesting aspect about silk fiber is that it is created under mild conditions in contrast to the severe processing conditions used to manufacture man-made strong fibers [224]. According to Lele et al. the magnificent strength of the spider fiber is attributed to strong hydrogen bonding between the protein chains [225]. Silk fibers differ in structure and properties depending upon their source [226]. Silks are characterized by secondary structure such as β-sheets which renders good mechanical properties to them. Because of the fact that we can now genetically manipulate the amino acid sequence of silk to achieve targeted properties, several researchers have attempted to exploit this material for various biomedical applications including tissue engineering. The protein engineering techniques have also been employed to scale up the production of silk and to also control its structural organization [227].

For musculoskeletal tissue engineering, scaffolds made out of silkworm silk have been widely used to tissue engineer anterior cruciate ligaments, bone, and cartilage [228–230]. Unlike many other scaffolds, silk scaffolds facilitate cell attachment and cell spreading without further modifications [231]. Recent studies using matrices made out of silk demonstrates that human MSCs (hMSC) adhere to them [230, 232]. Silk is known for its slow degradation kinetics and possesses biocompatible properties comparable to that of PGA and collagen [231]. The proteolytic degradation rate is mainly dependent upon the environmental conditions. Particularly, silk degrades completely within two years in vivo [231].

From the ongoing discussion it is clear that using natural polymers as tissue engineering scaffolds offer a wide range of advantages such as biological signaling, cell adhesion, cell responsive degradation and re-modeling. However, these materials lack adequate mechanical properties which compromise their utilization as unique scaffold materials. Another major concern of using natural materials is the possibility of immuno-rejection, and can also transfer various viruses even though proper screening and purification can overcome these limitations. To this end, various synthetic polymers have been specifically developed for tissue engineering scaffolds.

4.2
Synthetic Hydrogels

Recent years have witnessed a surge of interest in using synthetic hydrogels in tissue engineering approaches, and the most appealing factor about them is that their properties such as hydrophilic-hydrophobic balance, mechanical and structural properties, degradation profile, etc. can be molecularly tailored. Poly(ethylene oxide) (PEO), poly(ethylene glycol) diacrylate) (PEGDA), poly(vinyl alcohol) (PVA), poly(acrylic acid) (PAA), poly(acrylamidomtheyl propane sulfonic acid) (PAMPS), poly(hydroxyl ethyl methacrylate) (PHEMA), and poly(propylene fumerate-co-ethylene glycol) (see Fig. 5) are just a few examples of synthetic polymers used to create hydrogels and have found applications as tissue engineering scaffolds.

Poly(ethylene oxide) (PEO): PEO and its oligomer poly (ethylene glycol) (PEG) are widely used in various biomedical applications because of their biocompatibility, hydrophilicity, and resistance to protein adhesion and cell adhesion [123, 233]. It has been reported that PEG-modified proteins exhibit decreased immunogenicity and antigenicity at an increased circulation time in the body [234]. PEGDA with a molecular weight less than 20 000 Da can be dissolved in body fluid and can be eliminated from the body via excretion through the kidneys [233]. The resistance of PEG to protein adhesions has been widely used to create nonfouling surfaces, where PEG chains are immobilized onto the surface by covalent bonding or adsorption [9, 235]. The biocompatibility and hydrophilicity of PEG has been

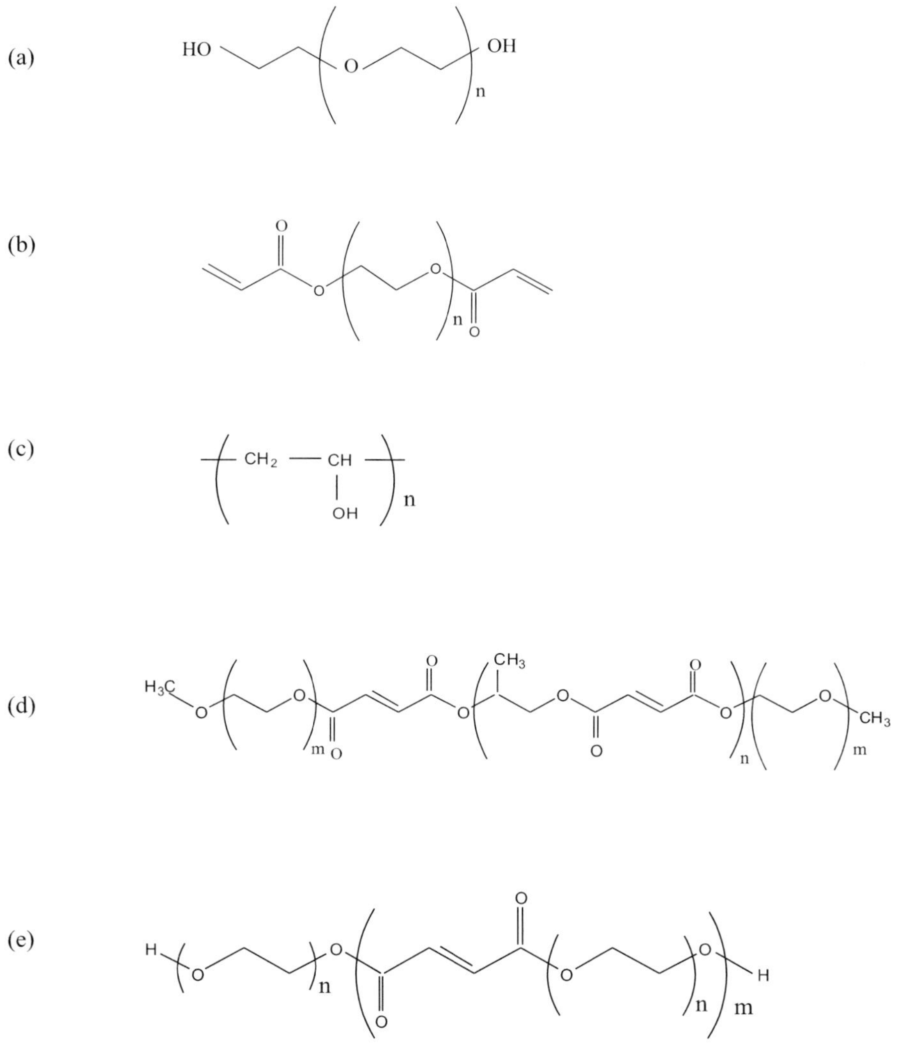

Fig. 5 Chemical structures of **a** poly(ethylene oxide), **b** poly(ethyleneglycol) diacrylate, **c** poly(vinyl alcohol), **d** poly(propylene fumerate-co-ethylene glycol, and **e** poly(ethylene glycol-fumerate)

explored to impart these properties on other materials by copolymerizing with PEG [236].

PEO hydrogels are created by γ-radiation or by UV photopolymerization when the PEO precursor contains acrylate termini such as PEGDA where the acrylate group functions as the crosslinkable moiety. Triblock copolymer of poly(ethylene oxide) and poly(propylene oxide), commercially known as

Pluronics, forms thermoreversible hydrogels in aqueous environments and has been used in various biomedical applications including tissue engineering scaffolds [237]. Our laboratory uses PEGDA scaffolds for cartilage engineering where photopolymerization of PEGDA is used to encapsulate chondrocytes, MSCs, and ES (Hwang et al, 2005, personal communication) [22, 24]. Our findings indicate that PEGDA serves as an efficient scaffold for anchorage-independent cells such as chondrocytes and also helps in tissue formation. Recently, our laboratory has also started exploring the possibility of using PEGDA hydrogel for guided regeneration of damaged cartilage (Casio et al, 2005, personal communication). We use micro-drilling to achieve migration of MSCs from the bone marrow to the defect site. Here, we hypothesize that the hydrogel will provide an adequate niche for the migrated MSCs to produce hyaline cartilage over fibrous cartilage. The nonadhesive nature of PEGDA hydrogels prevents the cells from adhering onto an unmodified scaffold. Hence, researchers have incorporated various peptides to impart adhesive properties to PEGDA [238]. Another major concern about the use of PEGDA is their nondegradable nature, which has been circumvented through the introduction of various degradable groups onto the polymer, although some degradation is observed when the material is in contact with cells or implanted [239, 240].

Poly(vinyl alcohol): PVA is a hydrophilic polymer and is generally prepared from poly(vinyl acetate) by hydrolysis, alcoholysis, or aminolysis [241]. PVA-based hydrogels have previously been used in various drug delivery devices such as artificial pancreas because of their biocompatibility [242]. Physical crosslinking of PVA can be achieved by repeated freeze-thawing of aqueous PVA solutions [243, 244]. The crystallites formed during the freeze-thawing process are responsible for creating the network [110]. The resulting network is stable and highly elastic at room temperature [245]. However, at higher temperature the crystallites melt, lose their stability, and dissolve in the solution. PVA can also be crosslinked chemically by using bi-functional crosslinking agents such as glutaraldehyde [104]. Boeckel et al. have used hydrogen bonds between PVA and amino acids to create novel cell-interactive hydrogels [9, 246]. Polyvinyl alcohol has also been modified with methacrylate groups to form photocrosslinkable polymers [133]. Unlike other hydrogels, PVA-based hydrogels possess good mechanical properties. Their mechanical properties have motivated many researchers to create PVA-based biomaterials for various biomedical applications such as artery, knee cartilage, and intervertebral disc replacements [245, 247–252]. Salubria is one such PVA-based biomaterial developed for artery and cartilage replacement.

Poly (fumerates): Biomaterials based on poly(propylene fumarate) (PPF) have been extensively used for orthopedic application such as injectable bone cement [253]. Unlike PEGDA and pluronics, fumerate-containing polymers are biodegradable since the ester link in the polymers can be cleaved hydrolytically. PPF is hydrophobic, and in order to synthesize hydrogels, it has been copolymerized with hydrophilic polymers such as poly(ethylene glycol) (PEG). The formed copolymer, poly(propylene fumarate)-co-poly(ethyleneglycol) is thermo-responsive, and hence it can be used in a minimally invasive manner [254]. In this copolymer system, the propylene fumarate repeat unit inherently contains a polymerizable vinyl group and a hydrolytically degradable ester group, while the poly (ethylene glycol) segments increases the hydrophilicity of the network so as to imbibe water. Poly(propylene fumarate)-co-poly(ethylene glycol) hydrogels have been reported to degrade in 12 weeks time both in vivo and in vitro [255]. All of these features make this polymer an attractive hydrogel scaffold for tissue engineering. Fisher et al. have explored the possibility of using this hydrogel for cartilage engineering [107].

Another fumarate-based hydrogel scaffold which has been studied for tissue engineering is poly(ethylene glycol fumarate) (OPF). OPF is synthesized by a reaction between poly(ethylene glycol) and fumaryl chloride. Temenoff et al. investigated the possibility of using OPF as a scaffold for bone tissue engineering [31, 43]. Their observations showed that the OPF hydrogels are suitable as an injectable cell carrier for bone and guided tissue regeneration.

5
Biodegradable Hydrogels

Most of the first generation hydrogel scaffolds are successful in providing structural support for the growth of the encapsulated cells in culture. However, only a few of the synthetic hydrogels are biodegradable. Biodegradability is an essential criterion for designing hydrogel scaffolds for tissue engineering applications. The hydrogels which initially provide a three-dimensional support to cells need to be degraded eventually as the cells differentiate and produce matrix. This is essential in order to maintain cell activities without any hindrance from the scaffold, as well as for creating the matrix with the desired cytoarchitecture. The kinetics of the degradation process needs to be tuned according to the adopted tissue engineering strategy as well as the targeted tissue and organ. For example, cells that proliferate fast and produce matrix rapidly require a fast degrading scaffold, whereas tissue structures that need stability and physical strength require a slow degrading scaffold [30, 230]. Ideally, the scaffold degradation rate should closely parallel the rate of ECM production. One of the important factors regarding the use of biodegradable hydrogel is that the degradation products must be nontoxic, and need to be eliminated rapidly by metabolic degradation or excretion by the kidney.

In addition to scaffolding, biodegradable hydrogels are also used to deliver various growth factors, which play a pivotal role in tissue development to increase the in vivo efficiency as well as the longevity of the proteins [62, 63]. In this application, the degradation profile i.e. the release of bio-active molecules should ideally match their uptake by the cells [63]. Common growth factors used in musculoskeletal regeneration include transforming growth factors (TGF-β) and bone morphogenetic proteins (BMP) [63]. In addition to their role in cell phenotype, growth factors also influence the cell viability, proliferation, differentiation, and matrix production by providing the necessary soluble signals to the cells [22, 60, 70, 71, 162, 256]. The growth factors can be incorporated directly into the scaffold through covalent bonding or by absorption.

5.1
Hydrolytically Degradable Hydrogels

Most of the synthetic hydrogels are degraded through hydrolysis of ester or amide linkages located within the polymer backbone [240, 257, 258]. The hydrolytically susceptible groups can either be inherently a part of the polymer or may be externally incorporated onto the polymer scaffold [239, 259]. This is achieved by introducing susceptible chemical groups such as esters onto the existing polymer back bone [239], or by copolymerizing, blending, or grafting degradable oligomers with nondegradable polymers [260, 261]. Here, the hydrogel degradation kinetics can be tuned by varying the scaffold chemistry as well as the molecular weight of the polymer segments. Various research groups have attempted to fine-tune hydrogel degradation.

Anseth et al. have created biodegradable PEG and PVA-based hydrogels by incorporating PLA moieties through copolymerization [260]. Wang et al. introduced a phosphate group into the PEGDA backbone by reacting PEG and phophoryl chloride [239], whereas Jo et al. introduced fumarate groups into the PEG backbone by reacting PEG with fumaryl chloride [259] to render PEG polymers hydrolytically degradable. The incorporation of phosphate groups into the PEGDA backbone allows one to create bio-interactive hydrogels for bone engineering, where the phosphate group facilitates the mineralization by directly interacting with the calcium ions. In all the studies discussed in this section, the degradation profile of the hydrogel has been manipulated via the choice of the degradable link, and not by the environment.

5.2
Enzymatically Degradable Hydrogels

Hydrogels made out of natural polymers can easily be degraded by enzymes, and this is one of their main advantages as potential scaffolding materials. Hence a marriage between such natural polymers and synthetic polymers can be creatively used to develop semi-synthetic degradable hydrogels [262, 263]. Recently, there has been a strong push towards incorporating enzyme-susceptible groups into synthetic polymers to render them enzymatically degradable [48, 264–267]. In these approaches, the rate of degradation depends upon both the susceptible bonds as well as the enzymatic concentration in the scaffold environment.

Lutolf et al. created a synthetic scaffold for bone grafts based on the biological recognition principle [48, 267]. Here, the authors have synthesized a PEG-based hydrogel scaffold using a matrix metalloproteinase (MMP) labile bi-functional peptide as a crosslinker, which is susceptible to cell-triggered proteolysis, i.e. the hydrogel degrades into soluble products upon exposure to cell-secreted MMPs. In this case, the crosslinker determines the hydrogel material properties by responding to the cell-secreted MMPs. In another study, Halstenberg et al. designed PEG-protein based biomaterial by grafting PEGDA onto an artificial protein created by a recombinant DNA approach [266]. Here, the protein was designed so as to perform cell-mediated degradation, while the PEGDA were used to create the network via photopolymerization.

6
Minimally Invasive Strategies

Advances in polymeric materials engineering offer new opportunities for minimally invasive surgeries (MIS), aimed at minimizing patient trauma and speeding up recovery. Many research and clinical studies have indeed focused on tissue engineering systems that can be injected in a noninvasive manner for use in arthroscopic surgery [66, 101, 268]. One attractive method to reduce the implantation invasiveness is to create hydrogels within the body via in situ polymerization. The advantage of in situ polymerization is that the functionalized oligomers (containing both insoluble and soluble bioactive molecules) can be injected into the defect site through small incisions and their subsequent polymerization enables a homogenous encapsulation of cells within the hydrogel. Since the fluidic precursors of the cell-hydrogel system can fill any irregular defect shapes, hydrogel-based scaffolds are highly suitable for treating defects which are not easily accessible, unless one adopts an invasive surgical procedure. The in situ polymerization also results in improved contact between the native tissue and hydrogel. The use of hydrogels also obviates many complications associated with residual toxic solvents since the hydrogels are water based. To this end, various in situ polymerization techniques/hydrogels such as photopolymerization, stimuli responsive polymers, shape memory polymers, and self-assembling peptide-based systems have been widely explored for minimally invasive applications.

6.1
Photopolymerization

In this approach, components of the scaffold along with viable cells are injected in a fluid state into the defect site arthroscopically, followed by subsequent polymerization within the defect site using a light source such as UV radiation. Figure 6 demonstrates such an arthroscopic immobilization of PEGDA hydrogel in a human cadaveric knee where the defect was made on the medial tibial condyle. Here, liquid bi-functional PEGDA solution along with the photoinitiator (Irgacure D2959) is injected into the defect site using a syringe and an optical fiber with a low-intensity UV light is used for polymerizing the reactants. The advantage of photogelation over other crosslinking techniques is the spatial and temporal control, as well as the fast curing rate obtained under physiological conditions at room temperature, which makes photogelation especially attractive for tissue engineering [101]. Photopolymerization has been used to encapsulate various cell types such as pancreatic islets [269], smooth muscle cells [270], osteoblasts [271], chondrocytes [24], MSCs [22], and ES (Hwang et al. 2005, personal communication).

(a)   (b)

Fig. 6 a Photograph showing the in situ photopolymerization of a hydrogel in a defect site and; **b** defects in the femoropatellar groove of a bovine knee are filled with the hydrogel after 5 mins of photopolymerization

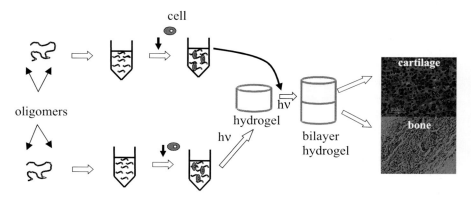

Fig. 7 Scheme for creating bilayer hydrogels for stratified tissues using photopolymerization

Photogelation is also a viable technique for controlling the spatial organization of different cell types within a three-dimensional system [24, 272]. This is accomplished by creating multilayer hydrogels where each layer is polymerized sequentially as shown in the schematic (Fig. 7). Integration between the layers is maintained while keeping cells confined within their respective layers. In our laboratory, such an approach has been used to encapsulate cells from different layers of the native cartilage within different hydrogel strips to create a stratified tissue engineered cartilage with zonal organization akin to the native one [24]. Additionally, multilayer hydrogels also enable the creation of other musculoskeletal tissues such as a whole knee joint with integrated cartilage and bone tissue [272].

6.2
Smart Hydrogels

Smart hydrogels, also known as stimuli-responsive hydrogels, are polymer networks that can undergo a discontinuous and macroscopic phase transition between a liquid and a solid state when subjected to a small change in one or more environmental stimuli such as temperature [105, 273], pH [105], light [274], radiation forces [275], and chemical triggers [115, 126–128, 276]. In the case of thermally reversible hydrogels, a subtle rise in the temperature near the lower critical solution temperature (LCST) can be used to trigger the sol-gel transition, where water-soluble coil structures transform into water-insoluble globular structures in aqueous medium, and vice versa. The LCST temperature is the manifestation of a critical balance of hydrophilic and hydrophobic groups, and significant research has been performed to tune the LCST temperature [105, 121].

One of the most extensively investigated stimuli responsive polymers is poly(N-isopropyl acrylamide), or in short, PNIPAm. PNIPAm exhibits a LCST temperature around $32\,^\circ$C, which means PNIPAm hydrogels undergo a reversible swelling-collapse volume phase transition at $32\,^\circ$C [105]. During this volume phase transition, the hydrogel expels the imbibed water. However, the phenomenon is reversible and the collapsed gel re-swells in water as the temperature is decreased below its LCST (i.e. $32\,^\circ$C). The re-swelling of the collapsed PNIPAm hydrogel as temperature decreases is accompanied by hysteresis, i.e. the threshold temperature at which the swollen phase transforms into a collapsed phase is different from the threshold temperature at which the reverse transition occurs. The transition temperature of thermoreversible polymers can be manipulated by altering the hydrophilic-hydrophobic balance [94]. For instance, copolymerizing PNIPAm with a hydrophobic monomer decreases its LCST temperature, whereas copolymerizing it with a hydrophilic monomer raises its LCST temperature [94].

In addition to PNIPAm, other synthetic thermoresponsive materials have been developed by copolymerizing hydrophilic and hydrophobic monomers or oligomers, such that the resulting copolymers show an LCST behavior at a certain critical composition of these reactants [105, 107]. The above phase separation behavior of polymers that especially have an LCST temperature close to the body temperature can be exploited to create tissue engineering scaffolds. Here, the in situ gelation of the starting polymer solution could be achieved by a subtle change in temperature [107]. Stile et al. used such PNIPAm-based hydrogels for cartilage tissue engineering as an injectable scaffold, which when gelled exhibited good chondrocyte cell viability [277]. In another study, peptide modified-PNIPAm hydrogels have been used to encapsulate rat calvarial osteoblasts with good cell viability [278]. Thermoreversible polypropylene fumarate-co-ethylene glycol hydrogels have also been utilized for bone tissue engineering as discussed earlier [107].

The temperature-triggered gelation of PNIPAm has also been explored for other biomedical applications such as drug delivery, and surface modification of cell culture dishes in order to create thermoresponsive culture surfaces upon which monolayers and multilayers of cells can be cultured and subsequently harvested noninvasively [279]. This technique known as "cell-sheet" engineering has been applied to a number of different cell types such as endothelial cells, kidney cells, and lung cells [280–283]. The latter application arises due to fact that at body temperature (37 °C), PNIPAm is hydrophobic and promotes cell adhesion, however at temperatures below 32 °C PNIPAm becomes hydrophilic, loses its cell adhesion properties and consequently releases the adhered cells. Therefore, one can easily lift off the cell sheets from the surface since this technique still retains cell-cell contacts. Such cell-sheets could be used to produce layered tissues by organizing the cell layers one on top of each other [282, 283]. However, a close look at the swelling-collapse phenomenon of PNIPAm shows that its hydration and dehydration kinetics are slow, and which can adversely affect the spontaneous cell-sheet detachment. In order to circumvent this limitation, Kwon et al. incorporated highly hydrophilic PEG oligomers at the interface between the PNIPAm chains and the cell layer [281].

Another approach to create a reversible biomimetic system is the creation of antigen responsive hydrogels [114, 115]. Miyata et al. developed a hydrogel, which can swell reversibly in response to a specific antigen (rabbit IgG) and change its structure. The hydrogel was prepared by grafting the antigen (rabbit IgG) and the corresponding antibody (GAR IgG) to the polymer network such that the binding between the antibody and antigen resulted in a crosslinked network. The hydrogel swells in the presence of free antigens because the physical crosslinks created by the antigen-antibody binding dissociate via the exchange of grafted antigens with the free antigens. Similar approaches can be used to create polymeric systems that could undergo in situ gelation for tissue engineering.

In addition to the above discussed synthetic polymers, some natural polymers such as gelatin, agarose, chitosan, etc, also undergo temperature-responsive gelation as mentioned earlier. The mechanism through which these materials undergo gelation is different from that corresponding to synthetic polymers. In these cases, denaturation of the triple helical confirmation in gelatin and double helical confirmation in polysaccharides results in a temperature-triggered gelation [284, 285]. Both gelatin and polysaccharides have been extensively investigated as tissue engineering scaffolds, as mentioned earlier [213, 284].

6.3
Self-Assembling Proteins

Another effective approach towards minimally invasive tissue engineering is via molecular self-assembly of proteins and protein-amphiphiles. The novelty of this approach is that the starting aqueous solution of proteins undergoes self-assembly-driven in situ gelation with respect to temperature, pH, and chemical triggers in the presence of biological fluids [117, 118, 286–290]. This self-assembly is generally mediated by secondary forces such as ionic interactions (e.g. electrostatic interactions), hydrogen bonds, hydrophobic interactions, and van der Waals interactions [291]. In addition to their in situ gelation capability, these protein-based hydrogels also provide the necessary biochemical cues to support cell proliferation and tissue formation [60, 287, 292].

Petka and coworkers creatively developed an artificial triblock protein, with a central hydrophilic peptide segment and terminal leucine zipper domains on both the ends, using recombinant DNA technology [118]. This protein undergoes reversible gelation with respect to changes in temperature and pH due to the dimerization and higher order aggregate formations of the leucine zipper motifs [118]. In another study, Wang et al. created another thermoreversible hydrogel hybrid containing both a synthetic polymer and an artificial protein segment where the two different synthetic polymers, N-(2-hydroxypropyl)-methacrylamide and N-(N9,N9-dicarboxymethylaminopropyl) methacrylamide, were crosslinked using coiled-coil protein-folding motifs [119].

Artificial elastin-like proteins (created by DNA recombinant technology) that undergo a reversible inverse phase separation were investigated as an injectable scaffold for cartilage tissue engineering by Betre and coworkers [293]. These elastin-like polypeptides enabled the encapsulated chondrocytes to retain their phenotype and produce extracellular matrix components of the cartilage such as sulfated glycosaminoglycan and collagen. Below their transition temperature, elastin-like polypeptides are soluble in water. Increasing the temperature leads to phase separation resulting in aggregation of polypeptides [293]. The LCST of elastin-like polypeptides can be controlled by varying the aminoacid sequence [294]. Urry and co-workers demonstrated that elastin-like polypeptides have mechanical properties reminiscent of native elastin, a protein found in muscle, ligaments, and cartilage [295]. In another study, Kisiday et al. investigated the effect of KLD-12 peptide hydrogel scaffolds on cartilage tissue engineering [286]. Both young and adult bovine chondrocytes seeded within these hydrogels were found to retain their characteristic phenotype and produced abundant glycosaminoglycans and type II collagen [286].

Self-assembly mediated in-situ gelation has also been utilized to create matrices with nanometer dimensions resembling the native ECM components [288, 296–299]. For example, Stupp and coworkers used supramolecular self-assembly to create nanofiber hydrogel scaffolds in situ by designing a peptide-amphiphile with a suitable balance of hydrophilic and hydrophobic groups [296–298]. In another study, the same group has developed in situ self-assembled nanofibrillar structures by utilizing electrostatic interactions between two oppositely charged peptide amphiphiles [299]. This method can also be applied to incorporate various bio-interactive ligands which promote cell adhesion into a single nanofiber by exploiting electrostatic interactions between peptides containing different amino acid sequences [288, 299].

6.4
Shape Memory Polymers

Another up-and-coming class of materials that could have a number of potential applications in minimally invasive surgical procedures involves shape memory polymers [127, 128, 300–303]. These materials have a huge potential application for many biomedical applications because of their unique properties, although, strictly speaking, they have not yet been applied to tissue engineering. Biomedical applications of shape memory materials hinge on the basic idea that these materials change their shape in response to certain external stimuli (such as temperature, pH, and light) from a compressed state, which can be administered through a noninvasive manner, to a bulky one, which can then fill the defect site. This application was motivated by metallic alloys such as NiTi (nitinol), which are known to exhibit shape memory (through martensitic structural transformation) and are well known for their applications in biomedical fields, e.g. surgical devices and implants [304]. However, polymer-based shape memory materials offer more advantages over such metal alloys, naturally due to their superior biocompatibility and biodegradability.

Polymers where reversible shape memory is induced by a change in temperature are known as thermo-responsive shape memory polymers. For example, a hydrogel formed by acrylic acid and stearyl acrylate shows significant temperature dependent mechanical properties [128]. Below 50 °C, this hydrogel behaves like a tough polymer whereas above 50 °C it behaves like a soft material. This transition allows one to process the hydrogels above 50 °C, where they are easily malleable, into the desired shape, which can be maintained at temperatures below 50 °C (such as room temperature). The hydrogel becomes soft and recovers its un-deformed original shape, which is predetermined by the covalent network, upon increasing the temperature above 50 °C [128]. Even though the concept is appealing, the use of this type of hydrogel is limited because the transition temperature is not commensurate with the body temperature.

Another class of hydrogels that demonstrate shape memory consists of a spatially modulated blend of two different hydrogels with different properties [302, 303]. For instance, when a bi-gel strip consists of temperature sensitive PNIPAm hydrogel and temperature insensitive polyacrylamide hydrogel, temperature increases causes it to bend. This change in shape is reversible and the gel can switch between the two shapes depending upon the external temperature. Such modulated hydrogels could be used to create various reversible shapes such as alphabets, numbers, wavy structures, and ribbon-like structures (refer to Fig. 8). Hydrogels of this nature could potentially be used to fill irregularly shaped defects in the biomedical field. However, one of the major concerns about these shape memory hydrogels is their weak mechanical properties and researchers are seeking to find tougher modulated polymers.

Recently, Lendlein et al. created a series of responsive shape memory polymers which are mechanically tough, biocompatible, and biodegradable, applicable to a number of biomedical applications [300, 301, 305]. Such shape memory polymers are achieved by copolymerizing precursors with different thermal characteristics such as the melting transition temperature. These shape memory polymers can be deformed into a temporary compressed state and they can recover the permanent shape only with the aid of an external stimulus such as temperature. This type of shape memory polymer mainly consists of two components: (i) molecular switches—precursors that can undergo stimuli responsive deformation and can fix the formed tempo-

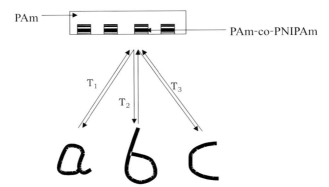

Fig. 8 Schematic of a shape memory modulated gel exhibiting different shapes depending upon the external stimuli

rary shape by physical crosslinks such as crystallites; and (ii) netpoints—precursors that determine the permanent shape of the polymer [301]. A biodegradable copolymer formed from oligo(ε-caprolactone)diol and oligo(p-dioxanone) diol has been shown to be an excellent suture material, as illustrated in Fig. 9. Here, the loose knots formed from fiber-like forms of this copolymer are used for wound closure where the knots tighten upon an increase in temperature, and thereafter keep the wound lips together under the right stresses for better healing [300].

Photoresponsive shape memory polymers are created by incorporating cinnamic groups which could be deformed to attain a temporary shape by UV light illumination at ambient temperatures. The un-deformed shape of these polymers can be recovered by exposing them to UV light with a different wavelength at ambient temperature [305]. The use of UV light as an external stimulus obviates the need for temperature changes, as required in the case of temperature-sensitive shape memory polymers.

Recently Thorton et al. designed chemically crosslinked alginate hydrogels into a compressed form for a noninvasive administration of the cells-scaffold system into the defect site. The inserted hydrogel is then swollen in vivo to re-gain its original equilibrium swollen shape [306]. Here, the initially compressed hydrogels regain their permanent shape through swelling and not through any external stimuli. This attempt therefore shows the feasibility of

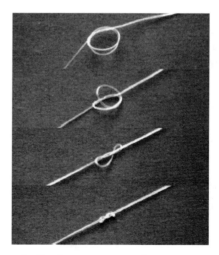

Fig. 9 A fiber of a thermoplastic shape-memory polymer was programmed by stretching about 200%. After forming a loose knot, both ends of the suture were fixed. The photo series shows, from *top* to *bottom*, how the knot tightened in 20 s when heated to 40 °C. (Reprinted with permission from Lendlein A, Langer R (2002) Science 269:1673, copyright 2002 AAAS)

using other more sophisticated shape memory polymers as discussed earlier in minimally invasive tissue engineering approaches.

6.5
Multifunctional Polymers

Another approach that has been employed for in situ hydrogel formation involves the mixing of two oligomer solutions having functional groups such that they form viscoelastic hydrogels upon reaction between the functional groups, similar to fibrin glue formation (Wang and Elisseeff, 2005, personal communication) [174, 307, 308]. To this end, Schiff base reaction of amine and aldehyde groups has been extensively explored in the past.

Balakrishnan et al. developed gelatin-based hydrogels in situ by mixing oxidized alginate and gelatin [307]. The hydrogels formed using this approach were nontoxic to the encapsulated mouse fibroblasts. In our laboratory, Wang and coworkers have used a similar approach for in situ integration of hydrogel implants (that can be immobilized onto the defect site) with the native tissue, which will be discussed in detail in Sect. 8 (Wang and Elisseeff, 2005, personal communication). The in situ gelation ability of two liquid reactants containing the functional groups, aldehyde functionalized CS and poly(vinyl alcohol-co-vinylamine), have been used for sealing corneal incisions by Reyes and coworkers [98]. Hyaluronic acid functionalized with various active groups such as amine and aldehyde groups can undergo in situ hydrogel formation when mixed with multifunctional monomers [174]. These materials are biocompatible, and in the presence of appropriate growth factors, these hydrogels can be used for bone or cartilage tissue engineering. Prestwich and coworkers used similar in situ gelation techniques to create HA and CS-based hydrogel films using poly(ethyleneglycol)-dialdehydes for wound healing applications [308]. These authors also applied in situ gelation of HA hydrogels for incorporating drugs that can be released in a controlled manner [263]. Sperinde and Griffith utilized enzyme-mediated crosslinking of peptide-modified PEG precursors to achieve in situ gelation. Such in situ formation of hydrogels can be used for tissue engineering and drug delivery applications without any toxicity concerns [99, 100]. This reaction enjoys the ability of transglutaminase to form amide linkage between the γ-carboxamide group of glutamine residues and primary amines such as the one in lysine. These covalent bonds between specific functional groups are strong and permanent unlike physical crosslinks.

7
Hydrogels for Stem Cell Differentiation

It is only in the last decade or so that the therapeutic value associated with donor/patient differentiated cells is being exploited to create functional replacements. Given the fact that the source of cells for repair is limited in such cases, it is desirable to exploit other cell lines. It is well documented in the literature that expansion of chondrocytes results in a loss of their phenotype as discussed earlier. Using biomaterial scaffolds and growth factors to re-differentiate the chondrocytes that have de-differentiated during monolayer expansions has had little success so far. For instance, Homicz et al. showed that only primary chondrocytes (immediately harvested prior to expansion) are suitable as seed cells for cartilage tissue engineering [309]. Chondrocytes that are used from subsequent passages produced decreased amounts of aggrecans and type II collagen as compared to those produced by primary cells. This therefore limits the mechanical as well as structural properties of the formed neocartilage [309]. Since chondrocytes lose their phenotype during expansions, it becomes necessary to start with a large biopsy to achieve enough cells for tissue engineering a cartilage sufficient in size for implantation. This leads to the primary constraint in using differentiated cells for cartilage tissue engineering, the lack in achieving adequate cell numbers. To circumvent the above constraint, scientists have begun to use stem cells (adult and embryonic) in tissue engineering for replacing dysfunctional tissues. This also avoids problems related to harvesting tissue as well as morbidity associated with the biopsy.

Nonhematopoietic stem cells (mesenchymal stem cells, MSCs) which are present in the bone marrow are multipotent, or in other words, they are capable of differentiating into various lineages such as cartilage [22, 310], bone [31, 311], connective tissues [312], etc. The differentiation of MSCs into a specific lineage requires a synergetic interaction between the cells and various insoluble and soluble growth factors during their culture [22, 313, 314]. This means that the tissue engineering scaffold needs to mimic the ECM components both structurally and functionally in order to provide the desired signals to direct cellular processes when using stem cells. Most of the first generation hydrogel scaffolds provides structural supports for the encapsulated cells but lack the required functional (biological) support. To increase the biofunctionality of the hydrogel scaffold one may covalently graft bioactive factors such as adhesion peptides or growth factors to the hydrogel [147, 315].

Micromass pellet cultures have shown the ability of MSCs to undergo chondrogenesis by providing enhanced cell-cell interactions [310]. However, to create organized tissues for treating large defects, support from three-dimensional scaffolds is required. Such three dimensional scaffolds accommodate a sufficient amount of cells, define the engineered tissue architecture, and may also be decorated with suitable biochemical cues to enhance the cell-matrix interactions. Hydrogels composed of various synthetic as well as natural polymers such as PEGDA, silk, collagen, etc., have been explored for the differentiation of MSCs into chondrocytes and osteoblasts [22, 230, 231]. In our laboratory, we investigated the ability of MSCs to undergo chondrogenesis both in vivo and in vitro (Sharma et al, 2005, personal communication) [22]. Our studies and others demonstrate the great potential of MSCs as an alternative cell source for cartilage tissue engineering, given the limitations of chondrocytes. In the case of in vivo studies, Sharma et al. used high molecular weight hyaluronan along with PEGDA oligomer solution to create a semi-interpenetrating network for the scaffold. In addition to increasing the viscosity of the starting solution for easy handling, the presence of HA also facilitates the differentiation of MSCs to chondrocytes. Moreover, the presence of HA in the hydrogel downregulates collagen type I expression indicating that HA plays a crucial role in modulating chondrocyte phenotype. These observations may be attributed to known biological functions of HA such as ligand-specific interactions with the cells. Hegewald et al. also observed a beneficial effect of HA on chondrogenic differentiation of equine MSCs [316].

In addition to their chondrogenic potential, MSCs have also been investigated for their ability to regenerate bone with the aid of three-dimensional hydrogel scaffolds. Recently, Meinel et al. investigated the effect of scaffold properties on differentiation of human bone marrow-derived mesenchymal stem cells (hMSC) into bone tissues by analyzing scaffolds made out of various materials such as collagen, crosslinked collagen, and silk [230]. According to these investigators, silk scaffolds outweigh the others in terms of performance. The authors attributed these observations to silk's high porosity, slow biodegradation, and structural integrity. These findings indicate that the differentiation of MSCs into the osteogenic lineage can be modulated by scaffold chemistry of the hydrogel systems. Temeoff et al. used hydrogels based on OPF for osteogenesis of MSCs. They observed enhanced osteogenesis within OPF hydrogels with higher swelling ability [31]. Wang et al. applied phosphoester-polyethylene glycol-based hydrogels for the encapsulation of MSCs and their subsequent osteogenic differentiation [311]. This hydrogel is hydrolytically degradable. The presence of phosphate groups in the hydrogel promotes the osteogenic differentiation of MSCs and autocalcification in addition to responding to the cell-secreted osteogenesis-related enzymes. The above studies demonstrate the requirement of suitable material properties and chemistry of scaffolds along with suitable growth factors and adequate degradation kinetics for the proper differentiation of MSCs.

Another cell line that could potentially be used for musculoskeletal tissue engineering is the embryonic stem (ES) cells. Embryonic stem cells are termed as pluripotent due to their versatile differentiation potential. However, controlled differentiation of ES cells into a particular lineage poses a grand challenge. Recently it has been reported that the encapsulation of ES cells within a scaffold system could increase its differentiation efficiency and allow formation of three-dimensional tissues (Hwang et al, 2005, personal communication) [72, 317, 318]. Guiding the differentiation of ES cells using a biomaterial scaffold is currently a relatively unexplored area, which has great potential in tissue engineering.

Hwang et al. compared the chondrogenic potential of ES cells in three-dimensional environments by encapsulating them within a PEGDA hydrogel scaffold with their monolayer culture (Hwang et al, 2005, personal communication). They observed enhanced chondrogenesis in the 3D system as compared to the conventional monolayer system. These findings further support the argument that the differentiation of ES can be modulated by 3D systems. Langer and coworkers also observed a similar trend in three-dimensional scaffolds leading to the formation of neural, hepatic, and cartilage tissues [72, 318]. They further investigated the role of mechanical properties of the scaffold on hES differentiation by using matrices such as matrigel, and a three-dimensional PLGA-co PLA scaffold coated with matrigel [72]. Although matrigel provides a three-dimensional environment to the encapsulated cells, it failed to support 3D tissue formation compared to the matrigel-coated PLGA-PLA scaffold. These findings indicate the importance of mechanical stiffness of the scaffold as well as the presence of biochemical cues in the scaffold on hES differentiation [72]. Philip et al. studied the effect of various ECM components on differentiation of rhesus ES cells. Their in vivo results show that the presence of cartrigel, an extract of cartilage matrix components, resulted in enhanced musculoskeletal differentiation of the ES cells [222]. These results indicate that the differentiation of ES cells can be directed to a particular lineage by using scaffolds containing ECM components derived from similar tissues. Xu et al. developed a feeder-free culture system for the proliferation of hES while retaining their undifferentiated state by exploiting the biological functions of Matrigel which is rich in ECM components [223].

The ongoing discussion clearly indicates the need for designing biomaterial scaffolds with desired biological cues and physical properties for the controlled differentiation of ES cells. To identify interactions between stem cells with a large array of biomaterials and growth factors, Anderson et al. applied high throughput microarray technology to screening biomaterials [319]. The advantage of this approach is that it enables the screening of large libraries of biomaterials in small quantities. Their findings indicate that certain biomaterials significantly induce the differentiation of hES cells into epithelial-like cells, whereas other biomaterials support the differentiation only in the presence or absence of specific growth factors.

8
Chemical Integration of Engineered Tissue with Native Tissues

Biomaterials have played a significant role in creating tissue-engineered implants and moving the field of regenerative medicine forward. The hurdle of moving from in vitro design of tissue engineering systems to their in vivo application still presents a challenge due to the lack of integration between the implant and the host tissue. While biomaterials such as hydrogels may be able to integrate with soft tissues, mechanically dense tissues such as cartilage and bone present significant challenges for creating strong tissue-biomaterial interactions. Cartilage poses a particularly difficult challenge for biomaterial integration since it lacks the ability to self repair, and possesses a dense extracellular matrix that impedes cellular migration and tissue integration. Integration of cartilage tissue has been the focus of numerous recent investigations with many studies examining the integration of both the engineered and native cartilage with host tissue.

Jurgensen et al. investigated the efficacy of tissue transglutaminase as a biological glue for integrating two cartilage pieces together [320]. In our lab-

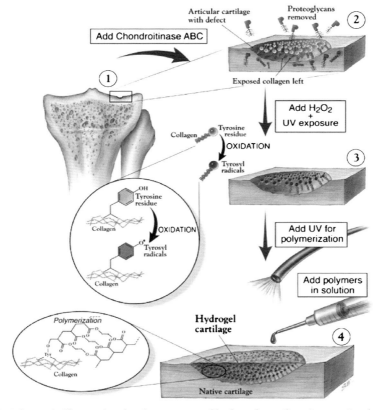

Fig. 10 Schematic illustration for the strategy of hydrogel-cartilage integration by tissue-initiated photopolymerization. Individual steps of the process are (1) clearance of the proteoglycans in cartilage by chondroitinase to expose the collagen network; (2) in situ generation of tyrosyl radicals by photo-oxidation of tyrosine residues on collagen with H_2O_2 under low intensity UV irradiation; and (3) introduction of a macromer solution and in situ photo-gelation via tyrosyl radical initiation and UV-excitation (Reprinted with permission Advanced Functional Materials [321])

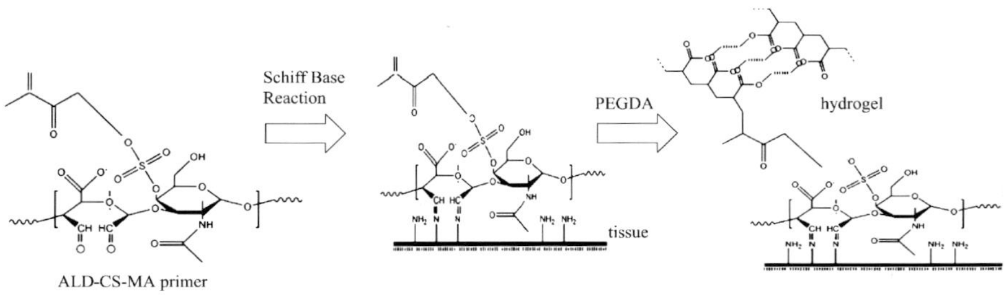

Fig. 11 Reaction scheme of hydrogel-tissue integration using ALD-CS-MA primer. The aldehyde groups react with amines in the cartilage tissue while methacrylate groups react with the polymerizing PEGDA precursor

oratory, we have exploited the ability of collagen chains located in the ECM to integrate covalently with the hydrogel implant via photopolymerization [321]. This technology, known as tissue-initiated photopolymerization, generates reactive tyrosyl radicals on the cartilage collagen backbone with the help of an oxidizing agent. The tyrosyl radicals on the cartilage, upon exposure to UV light, react with the oligomers (which also react amongst themselves to create a hydrogel) to create a well-integrated hydrogel-cartilage system, as shown in the schematic illustration (Fig. 10) [321].

As an alternative strategy, Wang et al. created a CS-based tissue adhesive to achieve covalent bond-mediated integration between the hydrogels and the host tissue. Here, CS is multi-functionalized with both methacrylate and aldehyde groups. The aldehyde groups react with collagen segments located within the host tissue whereas the methacrylate groups form covalent bonds between the adhesive and the PEGDA-cell solution during photopolymerization, which subsequently undergoes in situ gelation as shown in Fig. 11 (Wang and Elisseeff, 2005, personal communication). Our in vivo studies on the rabbit knee joint demonstrate the viability of this technique. In addition to anchoring the implant within the defect site initially, the strategy also enhances the integration of the engineered tissue with the native tissue at a molecular level by allowing migration of cells between the interface, and thereby regenerating tissue across the interface.

9
Concluding Remarks

This review outlined the progress over the past few decades in tissue engineering strategies that use hydrogels to treat musculoskeletal tissue loss. Significant advancements have been made in the design and development of hydrogel scaffolds which incorporate many of the required biochemical and biophysical cues for tissue development. Novel methods have been developed to create hydrogel scaffolds with excellent cell-adhesion properties and cell-mediated degradation profiles. Exciting new materials and strategies for in situ gelation of scaffold precursors have recently been formulated to enable minimally invasive application of materials to tissue defects. The application of stem cells, both adult and embryonic, to tissue repair strategies has made an impact in musculoskeletal tissue engineering. Preliminary results on differentiation of MSCs and ES cells into musculoskeletal lineages are very encouraging, thus pointing towards possible alternative sources of cells for tissue engineering. Future clinical practice for the treatment of musculoskeletal tissue loss will change and improve as these new technologies in biomaterial hydrogels and alternative cell sources are translated.

Acknowledgements We acknowledge Gaurav Arya (Department of Chemistry and Courant Institute of Mathematical Sciences, New York University) for valuable discussions. We also acknowledge financial support from the Whitaker Foundation and an arthritis investigator award from the Arthritis Foundation and Johns Hopkins University.

References

1. Haseltine WA (2003) The Brookings Review 21:38
2. Cho YR, Gosain AK (2004) Clin Plast Surg 31:377
3. Katti KS (2004) Colloid Surface B 39:133
4. Ambrose CG, Clanton TO (2004) Ann Biomed Eng 32:171
5. Langer R, Peppas NA (2003) AIChE J 49:2990
6. Chaikof EL, Matthew H, Kohn J, Mikos AG, Prestwich GD, Yip CM (2002) Ann NY Acad Sci 961:96
7. Lysaght MJ, Reyes J (2001) Tissue Eng 7:485
8. Gristina A (2004) Clin Orthop Relat R 427:4
9. Ratner BD, Bryant SJ (2004) Annu Rev Biomed Eng 6:41
10. Suh H, Park JC, Han DW, Lee DH, Han CD (2001) Artif Organs 25:14
11. Betz RR (2002) Orthopedics 25:S561
12. Hunziker EB (2001) Osteoarthr and Cartilage 10:432
13. Schreiber RE, Ratcliffe A (2000) Methods Mol Biol 139:301
14. Langer R, Vacnati JP (1993) Science 260:920
15. http://www.unos.org, date of visit: 05/15/05
16. Lysaght MJ, O'Loughlin JA (2000) ASAIOJ 46:515
17. http://www.bccresearch.com, date of visit: 05/15/05
18. Ma PX, Zhang R, Xiao G, Franceschi R (2001) J Biomed Mater Res 54:284
19. Alsberg E, Hill EE, Mooney DJ (2001) Crit Rev Oral Biol Med 12:64
20. An HS, Thonar EJMA, Masuda K (2003) Spine 28:S86
21. Hunter CJ, Matyas JR, Duncan NA (2003) Tissue Eng 9:667
22. Williams CG, Kim TK, Taboas A, Malik A, Manson P, Elisseef J (2003) Tissue Eng 9:679
23. Grande DA, Halberstadt C, Naughton G, Schwartz R, Manji R (1997) J Biomed Mater Res 34:211
24. Kim TK, Sharma B, Williams CG, Ruffner MA, Malik A, McFarland EG, Elisseeff JH (2003) Osteoarthr and Cartilage 11:653

25. Sah RL, Amiel D, Coutts RD (1995) Curr Opin Orthopedics 6:52
26. Chang SC, Tobias G, Roy AK, Vacanti CA, Bonassar LJ (2003) Plast Reconstr Surg 112:793
27. Temenoff JS, Mikos AG (2000) Biomaterials 21:431
28. Clavert JW, Weiss LE, Sundine MJ (2003) Clin Plast Surg 30:641
29. Rose FRAJ, Oreffo ROC (2002) Biochem and Biophys Res Commun 292:1
30. Alsberg E, Kong HJ, Hirano Y, Smith MK, Albeiruti A, Mooney DJ (2003) J Dent Res 82:903
31. Temenoff JS, Park H, Jabbari E, Sheffield TL, LeBaron RG, Ambrose CG, Mikos AG (2004) J Biomed Mater Res 70A:235
32. Zhnag J, Doll BA, Beckman EJ, Hollinger JO (2003) J Biomed Mater Res 67A:389
33. Kim BS, Nikolovski J, Bonadio J, Smiley E, Mooney DJ (1999) Exp Cell Res 251:318
34. Ma PX, Zhang R (2001) J Biomed Mater Res 56:469
35. Risbud MV, Albert TJ, Guttapalli A, Vresilovic EJ, Hillibrand AS, Vaccaro AR, Shapiro IM (2004) Spine 29:2627
36. Sakai D, Mochida J, Yamamoto Y, Nomura T, Okuma M, Nishimura K, Nakai T, Ando K, Hotta T (2003) Biomaterials 24:3531
37. Alini M, Li W, Markovic P, Aebi M, Spiro RC, Roughley PJ (2003) Spine 28:446
38. Yannas IV, Burke JF, Orgill DP, Skrabut EM (1982) Science 215:174
39. Machens HG, Berger AC, Mailaender P (2000) Cells Tissues Organs 167:88
40. Yannas IV, Lele E, Orgill DP, Skrabut EM, Murphy GF (1989) Proc Natl Acad Sci 86:933
41. Freyman TM, Yannas IV, Gibson LJ (2001) Prog Mater Sci 46:273
42. Ono I, Tateshita T, Inoue M (1999) J Biomed Mater Res 48:621
43. Temenoff JS, Steinbis ES, Mikos AG (2003) J Biomat Sci-Polymer E 14:989
44. Zellin G, Gritli-Linde A, Linde A (1995) Biomaterials 16:601
45. Fisher JP, Lalani Z, Bossano CM, Brey EM, Demian N, Johnston CM, Dean D, Jansen JA, Wong MEK, Mikos AG (2004) J Biomed Mater Res 68A:428
46. Shin H, Ruhe PQ, Mikos AG, Jansen JA (2003) Biomaterials 24:3201
47. Kellomaki M, Niiranen H, Puumanen K, Ashammakhi N, Waris T, Tormala P (2000) Biomaterials 21:2495
48. Lutolf MP, Weber FE, Schmoekel HG, Schense JC, Kohler T, Muller R, Hubbell JA (2003) Nat Biotechnol 21:513
49. Record RD, Hillegonds D, Simmons C, Tullius R, Rickey FA, Elmore D, Badylak SF (2001) Biomaterials 22:2653
50. Stephen B, Steven A, Pam P, Roger H, Vince M, Rhonda C, Christopher H (1999) Clin Orthop 367S:S333
51. Voytik-Harbin SL, Brightman AO, Kraine MR, Waisner B, Badylak SF (1997) J Cell Biochem 67:478
52. Aebischer P, Goddard M, Signore AP, Timpson RL (1994) Exp Neurol 126:151
53. Tessler A, Murray M (1996/1997) Neural Transplantation: Spinal Cord. In: Lanza RP, Chick WL (eds). Yearbook of Cell and Tissue Transplantation. Kluwer, Dordrecht, The Netherlands, p 175
54. Lanza RP, Chick WL (1997) Surgery 121:1
55. Sullivan SJ, Maki T, Borland KM, Mahoney MD, Solomon BA, Muller TE, Monaco AP, Chick WL (1991) Science 252:718
56. Pfeiffer N (1997) Genetic Eng News 17:1
57. Lars P, Thomas M, Mats B, Anders N, Eva SJ, Anders L (2000) Clin Orthop 374:212
58. Minas T (2001) Clin Orthop Relat R 391S:S349
59. http://www.carticel.com/home.asp, date of visit: 05/15/05
60. Lutolf MP, Hubbell JA (2005) Nat Biotechnol 23:47

61. Pavesio A, Abatangelo G, Borrione A, Brocchetta D, Hollander AP, Kon E, Torasso F, Zanasi S, Marcacci M (2003) Tissue Engineering of Cartilage and Bone. Novaris Foundation Symposium 249:203
62. Elisseeff J, McIntosh W, Fu K, Blunk T, Langer R (2001) J Orthopaed Res 19:1098
63. Babensee JE, McIntire LV, Mikos AG (2000) Pharm Res 17:497
64. Shea LD, Smiley E, Bonadio J, Mooney DJ (1999) Nat Biotechnol 17:551
65. Mahoney MJ, Saltzman WM (2001) Nat Biotechnol 19:934
66. Vacanti JP, Langer R (1999) Lancet 354:32
67. Vacanti CA, Langer R, Schloo B, Vacanti JP (1991) Plastic Reconstr Surg 88:733
68. Park YJ, Ku Y, Chung CP, Lee SJ (1998) J Contr Rel 51:201
69. Brekke JH, Toth JM (1998) J Biomed Mater Res 43:380
70. Mori M, Isobe M, Yamazaki Y, Ishihara K, Nakabayashi N (2000) J Biomed Mater Res 50:191
71. Saito N, Okada T, Horiuchi H, Murakami N, Takahashi J, Nawata M, Ota H, Nozaki K, Takaoka K (2001) Nat Biotechnol 19:332
72. Levenberg S, Huang NF, Lavik E, Rogers AB, Itskovitz-Eldor J, Langer R (2003) Proc Natl Acad Sci USA 100:12741
73. Boestman O (1990) J Bone Joint Surg 73B:592
74. Muggli DS, Burkoth AK, Anseth KS (1999) J Biomed Mater Res 46:271
75. Mathiowitz E, Jacob JS, Jong YS, Carino GP, Chickering DE, Chaturvedi P, Santos CA, Vijayaraghavan K, Montgomery S, Bassett M, Morrell (1997) Nature 386:410
76. Dang WB, Daviau T, Ying P, Zhao Y, Nowotnik D, Clow CS, Tyler B, Berm H (1996) J Contr Rel 42:83
77. Lavik E, Langer R (2004) Appl Microbiol Biot 65:1
78. Schmeltzer RC, Anastasiou TJ, Uhrich KE (2003) Polym Bull 49:441
79. Ibim SEM, Uhrich KE, Attawia M, Shastri VR, El-Amin SF, Bronson R, Langer R, Laurencin CT (1998) J Biomed Mater Res 43:374
80. Anseth KS, Shastri VR, Langer R (1999) Nat Biotechnol 17:156
81. Burdick JA, Anseth KS (2002) Biomaterials 23:4315
82. Cukierman E, Pankov R, Stevens DR, Yamada KM (2001) Science 294:1708
83. Grinnell F (2003) Trends Cell Biol 13:264
84. Chen CS, Yannas IV, Spectar M (1995) Biomaterials 16:777
85. Shastri VP, Martin I, Langer R (1999) Proc Natl Acad Sci USA 29:1970
86. Cooper AI (1997) J Mater Chem 10:207
87. Harris LD, Kim BS, Mooney DJ (1998) J Biomed Mater Res 42:396
88. Mooney DJ, Baldwin DF, Suh NP, Vacanti JP, Langer R (1996) Biomaterials 17:1417
89. Lu L, Peter SJ, Lyman MD, Lai HL, Leite SM, Tamada JA, Uyama S, Vacanti JP, Langer R, Mikos AG (2000) Biomaterials 21:1837
90. Freed LE, Vunjak-Novakovic G, Biron RJ, Eagles DB, Lesnoy DC, Barlow SK, Langer R (1994) Biotechnology (NY) 12:689
91. Hu Y, Grainger DW, Winn SR, Hollinger JO (2001) J Biomed Mater Res 59:563
92. Niklason LE, Gao J, Abbott WM, Hirschi KK, Houser S, Marini R, Langer R (1999) Science 284:489
93. Cao Y, Vacanti JP, Paige KT, Upton J, Vacanti CA (1997) Plast Reconstr Surg 100:297
94. Badiger MV, Lele AK, Bhalerao VS, Varghese S, Mashelkar R (1998) J Chem Phys 109:1175
95. Hennink WE, van Nostrum CF (2002) Adv Drug Del Rev 54:13
96. Dai WS, Barabari TA (1999) J Membr Sci 156:67
97. Yamamoto M, Tabata Y, Hong L, Miyamoto S, Hashimoto N, Ikada Y (2000) J Control Rel 64:133

98. Reyes JMG, Herretes S, Pirouzmanesh A, Wang DA, Elisseeff JH, Jun A, McDonnell PJ, Chuck RS, Behrens A (2005) Invest Ophth Vis Sci 46:1247
99. Sperinde JJ, Griffith LG (1997) Macromolecules 30:5255
100. Sperinde JJ, Griffith LG (2000) Macromolecules 33:5476
101. Elisseeff J, Anseth K, Sims D, McIntosh W, Randolph M, Langer R (1999) Proc Natl Acad Sci USA 96:3104
102. Bhalerao VS, Varghese S, Lele AK, Badiger MV (1998) Polymer 39:2255
103. Varghese S, Lele AK, Srinivas D, Mashelkar RA (1999) J Phys Chem B 103:9530
104. Durry JL, Mooney DJ (2003) Biomaterials 24:4337
105. Varghese S, Lele AK, Mashelkar RA (2000) J Chem Phys 112:3063
106. Jeong B, Bae YH, Kim SW (1999) Macromolecules 32:7064
107. Fisher JP, Jo S, Mikos AG, Reddi H (2004) J Biomed Mater Res 71A:268
108. Mata J, Joshi T, Varade D, Ghosh G, Bahadur P (2004) Colloid Surface A 247:1
109. Cascone MG, Lazzeri L, Sparvoli E, Scatena M, Serino LP, Danti S (2004) J Mater Sci-Mater M 15:1309
110. Ricciardi R, Auriemma F, Gaillet C, Rosa CD, Laupretre F (2004) Macromolecules 37:9510
111. Stenekes RJH, Talsma H, Hennink WE (2001) Biomaterials 22:1891
112. Bell CL, Peppas NA (1996) J Control Release 39:201
113. Mathur AM, Hammonds KF, Klier J, Scanton AB (1998) J Control Release 54:177
114. Miyata T, Asami N, Uragami T (1999) Macromolecules 32:2082
115. Miyata T, Asami N, Uragami T (1999) Nature 399:766
116. de Jong SJ, De Smedt SC, Wahls MWC, Demeester J, Kettenes-van den Bosch JJ, Hennink WE (2000) Macromolecules 33:3680
117. Cappello J, Crissman JW, Crissman M, Ferrari FA, Textor G, Wallis O, Whitledge JR, Zhou X, Burman D, Aukerman L, Stedronsky ER (1998) J Control Release 53:105
118. Petka W, Harden J, McGrath K, Wirtz D, Tirrell DA (1998) Science 281:389
119. Wang C, Stewart RJ, Kopecek J (1999) Nature 397:417
120. Zhang S, Marini DM, Hwang W, Santoso S (2002) Curr Opin Chem Biol 6:865
121. Lele AK, Badiger MV, Hirve MM (1995) Chem Eng News 22:3535
122. Dagani R (1997) Chem and Eng News June: 9
123. Peppas NA, Huang Y, Torres-Lugo M, Ward JH, Zhang J (2000) Annu Rev Biomed Eng 2:9
124. Hoffman AS (2001) Ann NY Acad Sci 944:62
125. Holtz JH, Asher SA (1997) Nature 389:829
126. Kataoka K, Miyazaki H, Bunya M, Okano T, Sakurai Y (1998) J Am Chem Soc 120:12694
127. Varghese S, Lele AK, Srinivas D, Sastry M, Mashelkar RA (2001) Adv Mater 13:1544
128. Osada Y, Matsuda A (1995) Nature 376:219
129. Chen GH, Hoffman AS (1995) Nature 373:49
130. Varghese S, Lele AK, Mashelkar RA (2006) J Polym Sci, Polym Chem 44:666
131. Smetana K (1993) Biomaterials 14:1046
132. Hoffman AS (2002) Adv Drug Del Rev 43:3
133. Bryant SJ, Anseth KS, Lee DA, Bader DL (2004) J Orthop Res 22:1143
134. Miot S, Woodfield T, Daniels AU, Suetterlin R, Peterschmitt I, Heberer M, van Blitterswijk CA, Riesle J, Martin I (2005) Biomaterials 26:2479
135. Grant GT, Morris ER, Rees DA, Smith PJ, Thom D (1973) FEBS Lett 32:195
136. Kong HJ, Alsberg E, Kaigler D, Lee KY, Mooney DJ (2004) Adv Mater 16:1917
137. Bouhadir KH, Lee KY, Alsberg E, Damm KL, Anderson KW, Mooney DJ (2001) Biotech Prog 17:945

138. Lee KY, Rowley JA, Eiselt P, Moy EM, Bouhadir KH, Mooney DJ (2000) Macromolecules 33:491
139. Eiselt P, Lee KY, Mooney DJ (1999) Macromolecules 32:5561
140. Grunder T, Gaissmaier C, Fritz J, Stoop R, Hortschansky P, Mollenhauer J, Aicher WK (2004) Osteoarthr Cartilage 12:559
141. Marijnissen W, van Osch G, Aigner J, van der Veen SW, Hollander AP, Verwoerd-Verhoef HL, Verhaar JAN (2002) Biomaterials 23:1511
142. Paige KT, Cima LG, Yaremchuk MJ, Schloo BL, Vacanti JP, Charles A (1996) Plast Reconstr Sur 97:168
143. Selden C, Hodgson H (2004) Transpl Immunol 12:273
144. Doyle JW, Roth TP, Smith RM, Li YQ, Dunn RM (1996) J Biomed Mater Res 32:561
145. Lin H, Yeh Y (2004) J Biomed Mater Res Part B: Appl Biomater 71:52
146. Awad HA, Wickham MQ, Leddy HA, Gimble JM, Guilak F (2004) Biomaterials 25:3211
147. Rowley JA, Madlambayan G, Mooney DJ (1999) Biomaterials 20:45
148. Diduch DR, Jordan LC, Mierisch CM, Balian G (2000) Arthroscopy 16:571
149. Nerem RM, Seliktar D (2001) Annu Rev Biomed Eng 3:225
150. Moreira PL, An YH, Santos AR, Genari SC (2004) J Biomed Mater Res B 71:229
151. Toolan BC, Frenkel SR, Pachence JM, Yalowitz L, Alexander H (1996) J Biomed Mater Res 31:273
152. Frenkel SR, Toolan B, Menche D, Pitman MI, Pachence JM (1997) J Bone Joint Surg 79B:831
153. Wiesmann HP, Nazer N, Klatt C, Szuwart T, Meyer U (2003) J Oral Maxil Surg 61:1455
154. Rammelt S, Schulze E, Bernhradt R, Hanisch U, Scharnweber D, Worch H, Zwipp H, Biewener A (2004) J Orthopaed Res 22:1025
155. Nehrer S, Breinan HA, Ramappa A, Shortkroff S, Young G, Minas T, Sledge CB, Yannas IV, Spector M (1997) J Biomed Mater Res 38:95
156. Buma P, Pieper JS, van Tienen T, van Susante JLC, van der Kraan PM, Veerkamp JH, van den Berg WB, Veth RPH, van Kuppevelt TH (2003) Biomaterials 24:3255
157. Dorotka R, Windberger U, Macfelda K, Bindreiter U, Toma C, Nehrer S (2005) Biomaterials 26:3617
158. Schuman L, Buma P, Versleyen D, de Man B, van der Kraan PM, van den Berg WB, Homminga GN (1995) Biomaterials 16:809
159. Ohno T, Tanisaka K, Hiraoka Y, Ushida T, Tamaki T, Tateishi T (2004) Mater Sci Eng C-Biomim 24:407
160. Qui W, Scully S (1996) Proc Of ORS, Atlanta, GA: 308
161. Lindsey WH, Ogle RC, Morgan RF, Cantrell RW, Sweeney TM (1996) Arch Otolaryngol-Head & Neck Surg 122:37
162. Hollinger JO, Winn SR (1999) Ann NY Acad Sci 875:379
163. Ber S, Kose GT, Hasirci (2005) Biomaterials 26:1977
164. Orban JM, Wilson LB, Kofroth JA, El-Kurdi MS, Maul TM, Vorp DA (2004) J Biomed Mater Res A 68:756
165. Rault I, Frei V, Herbage D, AbdulMalak N, Huc A (1996) J Mater Sci-Mater M 7:215
166. Zhao XB, Fraser JE, Alexander C, Lockett C, White BJ (2002) J Mater Sci-Mater M 13:11
167. Goldberg VM, Buckwalter JA (2005) Osteoarthr Cartilage 13:216
168. Jia X, Burdick JA, Kobler J, Clifton RJ, Rosowski JJ, Zeitels SM, Langer R (2004) Macromolecules 37:3239
169. Uebelhart D, Williams JM (1999) Curr Opin Rheumatol 11:427

170. Ramamurthy A, Vesely I (2003) J Biomed Mater Res 66A:317
171. Chen WYJ, Abatangelo G. Wound Rep Reg 7:79
172. Leach JB, Bivens KA, Patrick CW, Schmidt CE (2003) Biotechnol Bioeng 82:578
173. Kirker KR, Luo Y, Nielson JH, Shelby J, Prestwich GD (2002) Biomaterials 23:3661
174. Bulpitt P, Aeschlimann Y (1999) J Biomed Mater Res 47:152
175. Rhee WM, Berg RA (1996) US Patent 5 510 121
176. Leach JB, Bivens KA, Collins CN, Schmidt CE (2004) J Biomed Mater Res 70A:74
177. Aigner J, Tegeler J, Hutzler P, Campoccia D, Pavesio A, Hammer C, Kastenbauer E, Naumann A (1998) J Biomed Mater Res 42:172
178. Chow G, Knudson CB, Homandberg G, Knudson W (1995) J Biol Chem 270:27734
179. Larsen NE, Lombard KM, Parent EG, Balazs EA (1992) J Orthop Res 10:23
180. Yoo HS, Lee EA, Yoon JJ, Park TG (2005) Biomaterials 26:1925
181. Lindenhayn K, Perka C, Spitzer RS, Heilmann HH, Pommerening K, Mennicke J, Sit M (1999) J Biomed Mater Res 44:149
182. Grigo B, Roseti L, Fiorini M, Fini M, Giavaresi G, Aldini NN, Giardino R, Facchini A (2001) Biomaterials 22:2417
183. Park YD, Tirelli N, Hubbell JA (2003) Biomaterials 24:893
184. Bryant S, Davis-Arehart KA, Luo N, Shoemaker RK, Arthur JA, Anseth KS (2004) Macromolecules 37:6726
185. Li Q, Williams CG, Sun DN, Wang J, Leong K, Elisseeff JH (2004) J Biomed Mater Res 68A:28
186. Pieper JS, Osterhof A, Dijkstra PJ, Veerkamp JH, van Kuppevelt TH (1999) Biomaterials 20:847
187. Sechriest VF, Miao YJ, Niyibizi C, Larson AW, Matthew HW, Evans CH, Fu FH, Suh JK (2000) J Biomed Mater Res 49:534
188. Chang C-H, Liu HC, Lin CC, Chou CH, Lin FH (2003) Biomaterials 24:4853
189. Kirker KR, Luo Y, Morris SE, Shelby J, Prestwich GD (2004) J Burn Care Rehabil 25:276
190. Karp JM, Sarraf F, Shoichet MS, Davies JE (2004) J Biomed Mater Res 71A:162
191. Cox S, Cole M, Tawil B (2004) Tissue Eng 10:942
192. Peretti GM, Zaporojan V, Spangenberg KM, Randolph MA, Fellers J, Bonassar LJ (2003) J Biomed Mater Res 64A:517
193. Jockenhoevel S, Zund G, Hoerstrup SP, Chalabi K, Sachweh JS, Demircan L, Messmer BJ, Turina M (2001) Eur J Cardio-thorac 19:424
194. Ye Q, Zund G, Benedikt P, Jockenhoevel S, Hoerstrup SP, Sakyama S, Hubbell JA, Turina M (2000) Eur J Cardio-thorac 17:587
195. Westreich R, Kaufman M, Gannon P, Lawson W (2004) Laryngoscope 114:2154
196. Ting V, Sims CD, Brecht LE, McCarthy JG, Kasabian AK, Connelly PR, Elisseeff J, Gittes GK, Longaker MT (1998) Ann Plast Surg 40:413
197. Sims CD, Butler PEM, Cao YL (1998) Plast Reconstr Surg 101:1580
198. Perka C, Spitzer RS, Lindenhyan K, Sittinger M, Schultz O (2000) J Biomed Mater Res 49:305
199. Silverman R, Bonasser L, Passaretti D, Randolf MA, Yaremchuk MJ (1999) Plast Reconstr Surg 105:1393
200. Wechselberger G, Russell RC, Neumeister MW, Schoeller T, Piza-Katzer H, Rainer C (2002) Plast Reconstr Surg 110:123
201. Wechselberger G, Schoeller T, Arnulf S, Ninkovic M, Lille S, Russell RC (1998) J Urol 160:583
202. Horch RE, Bannasch H, Stark GB (2001) Transplant P 33:642
203. Currie LJ, Sharpe JR, Martin R (2001) Plast Reconstr Surg 108:1713

204. Chenite A, Chaput C, Wang D, Combes C, Buschmann MD, Hoemann CD, Leroux JC, Atkinson BL, Binette F, Selmani A (2000) Biomaterials 21:2155
205. Hejazi R, Amiji M (2003) J Control Release 89:151
206. Azad AK, Sermsintham N, Chandrkrachang S, Stevens WF (2004) J Biomed Mater Res 69B:216
207. Ishihara M (2002) Trends Glycosci Glyc 14:331
208. Mi FL, Wu YB, Shyu SS, Schoung JY, Huang YB, Tsai YH, Hao JY (2002) J Biomed Mater Res 59:438
209. Madihally SV, Matthew HW (1999) Biomaterials 20:1133
210. Tuzlakoglu K, Alves CM, Mano JF, Reis RL (2004) Macromol Biosci 4:811
211. Lee JE, Kim SE, Kwon IC, Ahn HJ, Cho H, Lee SH, Kin HJ, Seong SC, Lee MC (2004) Artificial Organs 28:829
212. Lahiji A, Sohrabi A, Hungerford DS, Frondoza CG (2000) J Biomed Mater Res 51:586
213. Francis Suh JK, Matthew HWT (2000) Biomaterials 21:2589
214. Seol YJ, Lee JY, Park YJ, Lee YM, Ku Y, Rhyu IC, Lee SJ, Han SB, Chung CP (2004) Biotechnol Lett 26:1037
215. Lu JX, Prudhommeaux, Meunier A, Sedel L, Guillemin G (1999) Biomaterials 10:1937
216. Hoemann CD, Sun J, Legare A, McKee MD, Buschmann MD (2005) Osteoarthr Cartilage 13:318
217. Hoemann CD, Hurtig M, Sun J, Rossomacha E, Shive MS, McKee MD, Chevrier A, Buschmann MD (2005) 50th Annual Meeting of the Orthopedic Research Society, Washington, DC
218. Muzzarelli RAA, Mattioli-Belmonte M, Tietz C, Biagini R, Ferioli G, Brunelli MA, Fini M, Giardino R, Ilari P, Biagini G (1994) Biomaterials 15:1075
219. Ono K, Saito Y, Yura H, Ishikawa K, Kurita A, Akaile T, Ishihara M (2000) J Biomed Mater Res 49:289
220. Nishimura K, Nishimura S, Nishi N, Saiki I (1984) Vaccine 2:93
221. Levengberg S, Langer R (2004) Curr Top Dev Biol 61:113
222. Philip D, Chen SS, Fitzgerald W, Orenstien J, Margolis L, Kleinman HK (2005) Stem Cells 23:288
223. Xu C, Inokuma MS, Denham J, Golds K, Kundu P, Gold JD, Carpenter MK (2001) Nat Biotechnol 19:971
224. Kitagawa M, Kitayama T (1997) J Mater Sci 32:2005
225. Lele AK, Joshi YM, Mashelkar RA (2001) Chem Eng Sci 56:5793
226. Altman GH, Diaz F, Jakuba C, Calabro T, Horan TL, Chen J, Lu H, Richmond J, Kaplan DL (2003) Biomaterials 24:401
227. van Hest JCM, Tirrell DA (2001) Chem Commun 1897
228. Sofia S, McCarthy MB, Gronowicz G, Kaplan DL (2001) J Biomed Mater Res 54:139
229. Altman GH, Horan RL, Lu HH, Moreau J, Martin I, Richmond JC, Kaplan DL (2002) Biomaterials 23:4131
230. Meinel L, Karageorgiou V, Hoffmann S, Fajardo R, Snyder B, Li C, Zichner L, Langer R, Vunjak-Novakovik G, Kaplan DA (2004) J Biomed Mater Res 71A:25
231. Vunjak-Novakovic G, Altman G, Horan R, Kaplan DL (2004) Annu Rev Biomed Eng 6:131
232. Meinel L, Hoffmann S, Karageorgiou V, Zichner L, Langer R, Kaplan DA, Vunjak-Novakovik G (2004) Biotechnol Bioeng 88:379
233. Bailey FE, Koleske JV (1976) Poly(ethylene oxide). Academic Press, New York, 6:103
234. Fuertges F, Abuchowski A (1990) J Control Release 11:139
235. Dalsin JL, Hu BH, Lee BP, Messersmith PB (2003) J Am Chem Soc 125:4253
236. Kim BS, Hrkach JS, Langer R (2000) Biomaterials 21:259

237. Liu Y, Chen F, Liu W, Cui L, Shang Q, Xia W, Wang J, Cui Y, Yang G, Liu D, Wu J, Xu R, Buonocore SD, Cao Y (2002) Tissue Eng 8:709
238. Yang F, Williams CG, Wang D, Lee H, Manson PL, Elisseeff J (2005) Biomaterials 26(30):5991
239. Wang DA, Williams CG, Li QA, Sharma B, Elisseeff JH (2003) Biomaterials 22:3969
240. Metters AT, Anseth KS, Bowman CN (2000) Polymer 41:3993
241. Finch CA (1973) Poly(vinyl alcohol): Properties and Applications. Wiley, London
242. Ohgawara H, Hirotani S, Miyazaki J, Teraoka S (1998) Artif Organs 22:788
243. Cascone MG, Laus M, Ricci D, Sbarbati del Guerra R (1995) J Mater Sci: Mater M 6:71
244. Hassan CM, Peppas NA (2000) Adv Poly Sci 153:37
245. Thomas J, Gomes K, Lowman A, Marcolong M (2004) J Biomed Mater Res 69B:135
246. Boeckel MS, Perry J, Leber ER, Nair P, Ratner BD (2003) Polym Preprnits 44:677
247. Oka M, Ushio K, Kumar P, Ikeuchi K, Hyon SH, Nakamura T, Fujita H (2000) Proc Inst Mech Eng [H] 214:59
248. Kobayashi M, Toguchida J, Oka M (2003) Biomaterials 24:639
249. Yura S, Oka M, Ushio K, Fujita H, Nakamura T (1999) In: Stein H, Suk H, Leung PC (eds) SIROT 99. Freund, London, p 371
250. Covert RJ, Ott RD, Ku DN (2003) Wear 255:1064
251. Stammen JA, Williams S, Ku DN, Guldberg RE (2001) Biomaterials 22:799
252. Kobayashi M, Chang YS, Oka M (2005) Biomaterials 26:3243
253. Peter SJ, Kim P, Yasko AW, Yaszemski MJ, Mikos AG (1999) J Biomed Mater Res 44:314
254. Behravesh E, Shung AK, Jo S, Mikos AG (2002) Biomacromolecules 3:153
255. Suggs LJ, Krishnan RS, Garcia CA, Peter SJ, Anderson JM, Mikos AG (1998) J Biomed Mater Res 42:312
256. Bentz H, Schroeder JA, Estridge TD (1998) J Biomed Mater Res 39:539
257. He S, Yaszemski MJ, Yasko AW, Engel PS, Mikos AG (2000) Biomaterials 21:2389
258. Agrawal CM, Ray RB (2001) J Biomed Mater Res 55:141
259. Jo S, Shin H, Shung AK, Fisher JP, Mikos AG (2001) Macromolecules 34:2839
260. Anseth KS, Metters AT, Bryant SJ, Martens PJ, Elisseeff JH, Bowman CN (2002) J Control Rel 78:199
261. Taniguchi I, Mayes AM, Chan EWL, Griffith LG (2005) Macromolecules 38:216
262. van Dijk-Wolthuis WNE, Tsang SKY, Kettenes-van den Bosch, Hennik WE (1997) Polymer 38:6235
263. Luo Y, Kirker RK, Prestwich GD (2000) J Contr Release 69:169
264. West JL, Hubbell JA (1999) Macromolecules 32:241
265. Mann BK, Gobin AS, Tsai AT, Schmedlen RH, West JL (2001) Biomaterials 22:3045
266. Halstenberg S, Panitch A, Rizzi S, Hall H, Hubbell JA (2002) Biomacromolecules 3:710
267. Lutolf MP, Raeber GP, Zisch AH, Tirelli N, Hubbell JA (2003) Adv Mater 15:888
268. Fisher JP, Dean D, Engel PS, Mikos AG (2001) Annu Rev Mater Res 31:171
269. Cruise GM, Hegre OD, Lamberti FV, Hager SR, Hill R, Scharo DS, Hubbell JA (1999) Cell Transplantation 8:293
270. Wong JY, Velasco A, Rajagopalan P, Pham Q (2003) Langmuir 19:1908
271. Burdick JA, Mason MN, Hinman AD, Thorne K, Anseth KS (2002) J Control Rel 83:53
272. Sharma B, Mountziaris PM, Khan M, Eliseeff JH (2005) International Symposium on Biomaterials and Biomechanics, Montreal, Quebec, Canada
273. Tanaka T (1978) Phys Rev Lett 40:820
274. Irie M, Yoshifumi M, Tusuyoshi T (1993) Polymer 34:4531

275. Juodkazis S, Mukai N, Wakaki R, Yamaguchi A, Matsuo S, Misawa H (2000) Nature 408:178
276. Kokufuta E, Zhang Y-Q, Tanaka T (1991) Nature 351:302
277. Stile RA, Burghardt WR, Healy KE (1999) Macromolecules 32:7370
278. Stile RA, Healy KE (2001) Biomacromolecules 2:185
279. Yang J, Yamato M, Okano T (2005) MRS bulletin 3:189
280. Kushida A, Yamato M, Kikuchi A, Okano T (2001) J Biomed Mater Res 54:37
281. Kwon OH, Kikuruchi A, Yamato M, Okano T (2003) Biomaterials 24:1223
282. Kwon OH, Kikuchi A, Yamato M, Sakurai Y, Okano T (2000) J Biomed Mater Res 50:82
283. Canavan HE, Cheng X, Graham DJ, Ratner BD, Castner DG (2005) Langmuir 21:1949
284. Lee CH, Singla A, Lee Y (2001) Int J Pharm 221:1
285. Jeong B, Kim SW, Bae YH (2002) Adv Drug Deliver Rev 54:37
286. Kisiday J, Jin M, Kurz B, Hung H, Semino C, Zhang S, Grodzinsky AJ (2002) Proc Natl Acad Sci USA 99:9996
287. Zhang S (2003) Nat Biotech 21:1171
288. Zhang S, Holmes TC, DiPersio CM, Hynes RO, Su X, Rich A (1995) Biomaterials 16:1385
289. Collier JH, Hu BH, Ruberti JW, Zhang J, Shum P, Thompson DH, Messersmith PB (2001) J Am Chem Soc 123:9463
290. Nowak AP, Breedveld V, Pakstis L, Ozbas B, Pine DJ, Pochan D, Deming TJ (2002) Nature 417:424
291. Varghese S (2001) PhD Thesis, Pune University, India
292. Holmes TC, Delacalle S, Su X, Rich A, Zhang S (2000) Proc Natl Acad Sci USA 97:6728
293. Betre H, Setton LA, Meyer DE, Chilkoti A (2002) Biomacromolecules 3:910
294. Meyer DE, Chilkoti A (1999) Nature Biotechnology 17:1112
295. Lee J, Macosko CW, Urry DW (2001) Macromolecules 34:5968
296. Hartgerink JD, Beniash E, Stupp SI (2001) Science 294:1684
297. Hartgerink JD, Beniash E, Stupp SI (2002) Proc Natl Acad Sci USA 99:5133
298. Silva GA, Czeisler C, Niece K, Beniash E, Harrington DA, Kessler JA, Stupp SA (2004) Science 303:1352
299. Niece KL, Hartgerink JD, Donners JJJM, Stupp SI (2003) J Am Chem Soc 125:7146
300. Lendlein A, Langer R (2003) Science 296:1673
301. Lendlein A, Schmidt AM, Langer R (2001) Proc Natl Acad Sci USA 98:842
302. Hu Z, Zhang X, Li Y (1995) Science 269:525
303. Li Y, Hu Z, Chen Y (1997) J Appl Polym Sci 63:1173
304. Lipscomb IP, Nokes LDM (1996) The application of shape memory alloys in medicine. Wiley, New York
305. Lendlein A, Jinag H, Junger O, Langer R (2005) Nature 434:879
306. Thornton AJ, Alsberg E, Albertelli M, Mooney DJ (2004) Transplantation 77:1798
307. Balakrishnan B, Jayakrishnan A (2005) Biomaterials 26:3941
308. Kirker KR, Prestwich GD (2004) J Polym Sci B Polym Phys 42:4344
309. Homicz MR, Schumacher BL, Sah RL, Watson D (2002) Otolaryngol Head Neck Surg 127:398
310. Solchaga LA, Goldberg VM, Caplan AI (2001) Clin Orthop Relat Res 391:S161
311. Wang DA, Williams CG, Yang F, Cher N, Lee H, Elisseeff JH (2005) Tissue Eng 11:201
312. Jiang Y, Jahagirdar BN, Reinhardt RL, Schwartz RE, Keene CD, Ortiz-Gonzalez XR, Reyes M, Lenvik T, Lund T, Blackstad M, Du J, Aldrich S, Lisberg A, Low WC, Largaespada DA, Verfaillie CM (2002) Nature 418:41

313. Pang Y, Cui P, Chen W, Gao P, Zhang H (2005) Arch Facial Plat Surg 7:7
314. Chen CP, Park Y, Rice KG (2004) J Peptide Rel 64:237
315. Shin H, Jo S, Mikos AG (2003) Biomaterials 24:4353
316. Hegewald AA, Ringe J, Bartel J, Kruger I, Notter M, Barnewitz D, Kaps C, Sittinger M (2004) Tissue & Cell 6:431
317. Liu H, Roy K (2005) Tissue Eng 11:319
318. Levenberg S, Golub JS, Amit M, Itskovitz-Eldor J, Langer R (2002) Proc Natl Acad Sci USA 99:4391
319. Anderson DG, Levenberg S, Langer R (2004) Nat Biotechnol 22:863
320. Jurgensen K, Aeschlimann D, Cavin V, Genge M, Hunziker EB, Switzerland B (1997) J Bone Joint Surgery-Am 79A:185
321. Wang DA, Williams CG, Yang F, Elisseeff JH (2004) Adv Funct Mater 14:1152

Adv Polym Sci (2006) 203: 145–170
DOI 10.1007/12_088
© Springer-Verlag Berlin Heidelberg 2006
Published online: 21 April 2006

Self-Assembling Nanopeptides Become a New Type of Biomaterial

Xiaojun Zhao[1] (✉) · Shuguang Zhang[2]

[1]Institute for NanoBiomedical Technology and Membrane Biology, West China Hospital,
Sichuan University, Research Building No 1, West China Hospital Science Park No 4,
Gao Peng Rd., 610041 Chengdu, China
xiaojunz@mit.edu

[2]Center for Biomedical Engineering, Center for Bits and Atoms,
Massachusetts Institute of Technology, 500 Technology Square,
Cambridge, MA 02139-4307, USA

Abstract Combining physics, engineering, chemistry and biology, we can now design, synthesize and fabricate biological nano-materials at the molecular scale using self-assembling peptide systems. These peptides have been used for fabrication of nanomaterials including nanofibers, nanotubes and vesicles, nanometer-thick surface coating and nanowires. Some of these peptides are used for stabilizing membrane proteins, and others provide a more permissive environment for cell growth, repair of tissues in regenerative medicine, and delivering genes and drugs. Self-assembling peptides are also useful for fabricating a wide spectrum of exquisitely fine architectures, new materials and nanodevices for nanobiotechnology and a variety of engineering. These systems lie at the interface between molecular biology, chemistry, materials science and engineering. Molecular self-assembly will harness nature's enormous power to benefit other disciplines and society.

Keywords Regenerative medicine · Polymers · Nanobiotechnology ·
Self-assembly peptide · Designer nanomaterials

1
Introduction

1.1
The Nature's Building Blocks at the Molecular Scale and Design, Synthesis and Fabrication

Nature is the grandmaster when it comes to building extraordinary materials and molecular machines—from the bottom up, one atom and one molecule at a time. Multifunctional macromolecular assemblies in living organisms, including hemoglobin, polymerases, ATP synthase, membrane channels, the spliceosome, the proteosome, ribosomes, and photosystems are all essentially exquisitely designed molecular machines acquired through billions of years of prebiotic molecular selection and evolution. Nature has produced a basic set of molecules that includes 20 amino acids, a few nucleotides, a dozen or so lipid molecules and few dozens of sugars as well as naturally modified building blocks or metabolic intermediates. With these seemingly simple molecules, natural processes are capable of fashioning an enormously diverse range of fabrication units, which can further self-organize into refined structures, materials and molecular machines that not only have high precision, flexibility and error correction, but also are self-sustaining and evolving. Indeed, nature favors bottom-up design, building up from molecular assemblies, bit by bit, more or less simultaneously in a well-defined manner.

1.2
A Fabrication Tool: Nanobiotechnology Through Molecular Self-Assembly

Design of molecular biological nanostructures requires detailed structural knowledge to build advanced materials and complex systems. Using basic biological building blocks and a large number of diverse peptide structural motifs [1, 2], it is possible to build new materials from the bottom up. Molecular self-assembly is ubiquitous in nature, from lipids that form oil droplets in water and surfactants that form micelles and other complex structures in water to sophisticated multiunit ribosome and virus assemblies. These systems lie at the interface of molecular and structural biology, protein science, chemistry, polymer science, materials science and engineering. Many self-assembling systems have been developed, which range from organic supramolecular systems, bi-, tri-block copolymers [3], and complex DNA structures [4, 5], simple and complex proteins [6–8] to peptides [9–22].

1.3
Basic Engineering Principles for Micro- and Nano-Fabrication
Based on Molecular Self-Assembly Phenomena

Programmed assembly and self-assembly are ubiquitous in nature at both macroscopic and microscopic scales. The Great Wall of China, the Pyramids of Egypt, the schools of fish in the ocean, flocks of birds in the sky, protein folding and oil droplets on water are all such examples. On the other hand, self-assembly describes the spontaneous association of numerous individual entities into a coherent organization and well-defined structures to maximize the benefit of the individual without external instruction. If we shrink construction units by many orders of magnitude into nano-scale, such as structurally well-ordered protein fragments, or peptides [21], we can apply similar principles to construct molecular materials and devices, through molecular self-assembly and programmed molecular assembly.

1.4
Both Chemical Complementarity and Structural Compatibility
for Bionanotechnology

The "bottom-up" approach, by which materials are assembled molecule by molecule (and in some cases even atom by atom) to produce novel supramolecular architectures is a powerful technology. This approach is likely to become an integral part of materials manufacture and requires a deep understanding of individual molecular building blocks and their structures, assembly properties and dynamic behaviors. Molecular self-assembly interactions typically include hydrogen bonds, electrostatic attractions, and Van der Waals interactions. Although these bonds are relatively insignificant in isolation, when combined together as a whole, they govern the structural conformation of all biological macromolecules and influence their interaction with other molecules. The water-mediated hydrogen bond is especially important for living systems, as all biological materials interact with water. It is a powerful approach for fabricating novel supramolecular architectures, which is ubiquitous in nature and has now emerged as a new approach in chemical synthesis, nanotechnology, polymer science, materials and engineering.

To date, several self-assembling peptide systems have been studied, ranging from models for studying protein folding and protein conformational diseases, to molecular materials for producing peptide nanofibers, peptide scaffolds, peptide surfactants and peptide ink [9, 10] easy to produce at a large scale to drive the development of this new industry. These self-assembly peptide systems represent a significant advance in molecular engineering for diverse technological innovations. This field is growing at a rapid pace and it is impossible to summarize all aspects of the work being done by others in this limited space, and hence this review focuses on a few examples especially from our laboratory. We focus on our work from the past decade, but those who are interested in trends over a longer period of time are referred to earlier reviews [10, 11].

2
Self-Assembly Peptide Systems

A new class of oligopeptide-based biological materials was serendipitously discovered from the self-assembly of ionic self-complementary oligopeptides [3]. A number of peptide molecular self-assembly systems has been designed and developed. This new class of biological materials has considerable potential for a number of applications, including scaffolding for tissue repair and tissue engineering, drug delivery of molecular medicine and biological surface engineering. Molecular self-assembly relies on chemical complementarity and structural compatibility [23]. These fundamentals are keys to the design of the molecular units required for the fabrication of functional macrostructures, which in turn permit molecular self-assembly in nanotechnology and nanobiotechnology.

The complementary ionic sides have been classified into several moduli (modulus I, modulus II, modulus III, modulus IV, etc., and mixtures thereof). This classification is based on the hydrophilic surfaces of the molecules, which have alternating positively and negatively charged amino acids alternating by one residue, two residues, and three residues and so on. For example, charge arrangements for modulus I, modulus II, modulus III and modulus IV are -+-+-+-+, --++--++, ---+++ and ----++++, respectively. The charge orientation can also be designed in the reverse orientation, which can yield entirely different molecules. These well-defined sequences allow the peptides to undergo ordered self-assembly, in a process resembling some situations found in well-studied polymer assemblies. A broad range of peptides and proteins have been shown to produce very stable nanofiber structures, also called amyloid fibers [24–34].

2.1
Peptides as Construction Motifs

Similar to the construction of a house, many other parts of the house, such as doors and windows can be prefabricated and program-assembled according to architectural plans. If we shrink the construction units many orders of magnitude to the nanoscale, we can apply similar principles for constructing molecular materials and devices, through molecular self-assembly and programmed molecular assembly.

2.2
Modulus I: "Peptide Lego"

Type I peptides, also called "molecular Lego" are the first member of the "peptide Lego", which was serendipitously discovered from a segment in a left-handed Z-DNA binding protein in yeast and named Zuotin [14]. Lego bricks have pegs and holes, which can be assembled into particular structures. In a similar way, these peptides can be assembled at the molecular level. The nanometer scale "peptide Lego" resembles Lego bricks that have both pegs and holes in a precisely determined organization and can be programmed to assemble into well-formed structures. This class of "peptide Lego" can spontaneously assemble into well-formed nanostructures at the molecular level [15].

The molecular structure and proposed complementary ionic pairings of the modulus I peptides between positively charged lysines and negatively charged glutamates in an overlapping arrangement are modeled in Fig. 1. This structure represents an example of this class of self-assembling β-sheet peptides that spontaneously undergo association under physiological conditions. If the charged residues are substituted, i.e. the positively charged lysines (Lys) are replaced by the positively charged arginines (Arg) and the negatively charged glutamates (Glu) were replaced by negatively charged aspartates (Asp), the peptide would still be able to undergo self-assembly into macroscopic materials. However, if the positively charged residues, Lys and Arg, were replaced by negatively charged residues, Asp and Glu, the peptide would not be able to undergo self-assembly and form macroscopic materials although β-sheet structures have been observed in the presence of salt. If the alanines (Ala) were changed to more hydrophobic residues, such as Leu, Ile, Phe or Tyr, the molecules had a greater tendency to self-assemble and formed

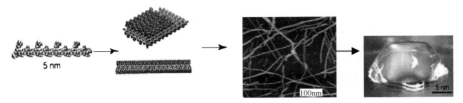

Fig. 1 Fabrication of various peptide materials. Peptide Lego, also called ionic self-complementary peptide has 16 amino acids, 5 nm in size, with an alternating polar and non-polar pattern. They form stable b-strand and b-sheet structures, thus the side chains partition into two sides, one polar and the other non-polar. They undergo self-assembly to form nanofibers with the non-polar residues inside (*green*) and positive (*blue*) and negative (*red*) charged residues forming complementary ionic interactions, like a checkerboard. These nanofibers form interwoven matrices that produce a scaffold hydrogel with very high water content, 99.5% water (images courtesy of Hidenori Yokoi)

peptide matrices with enhanced strength [35]. The fundamental design principles of such self-assembling peptide systems can be readily extended to polymers and polymer composites, where copolymers can be designed and produced.

2.3
Molecular Switches

Several peptides have been developed as "molecular switches" in which the peptides can drastically change their molecular structure. One of the peptides with 16 amino acids, DAR16-IV, has a β-sheet structure 5 nm in length at ambient temperature but can undergo an abrupt structural transition at high temperatures to form a stable α-helical structure 2.5 nm long [13]. Similar structural transformations can be induced by changes in pH. This suggests that secondary structures of some sequences, especially segments flanked by clusters of negative charges on the N-terminus and positive charges on the C-terminus, may undergo drastic conformational transformations under the appropriate conditions. These findings do not only provide insights into protein–protein interactions during protein folding and the pathogenesis of some protein conformational diseases, such as Alzheimer's disease, Gestmann–Straussler–Scheiker syndrome and/or kuru in humans and scrapie in sheep, cow, mink or elk, as well as certain types of cancer, all of which are examples of such conformational disorder [34–41], but can also be developed as molecular switches for a new generation of nanoactuators. Both peptides of DAR16-IV (DADADADARARARARA) and EAK12 (AEAEAEAEAKAK) have a cluster of negatively charged glutamate residues close to the N-terminus and a cluster of positively charged Arg residues near the C-terminus. It is well known that all α-helices have a helical dipole moment with a partially negative C-terminus toward a partially positive

N-terminus [42]. Because of the unique sequence of DAR16-IV and EAK12, their side chain charges balance the helical dipole moment, therefore favoring helical structure formation. However, they also have alternating hydrophilic and hydrophobic residues as well as ionic self-complementarity, which have been previously found to form stable β-sheets. Thus, the behavior of these Type II molecules is likely to be more complex and dynamic than other stable β-sheet peptides. Additional molecules with such dipoles have been designed and studied, and the results confirmed the initial findings. Others have also reported similar findings that proteins and peptides can undergo self-assembly and disassembly or change their conformations depending on the environmental influence, such as its location, pH change, and temperature, or crystal lattice packing [43–45].

2.4
Peptide Ink

"Peptide inks", undergo self-assembly on the surface rather than with themselves. They form monolayers on surfaces for a specific cell pattern formation or to interact with other molecules. These oligopeptides have three distinct features. The first feature is the terminal segment of ligands that incorporate a variety of functional groups for recognition by other molecules or cells. The second feature is the central linker where a variable spacer is not only used to allow freedom of interaction at a specified distance away from the surface but also controls the flexibility or rigidity. The third feature is the surface an-

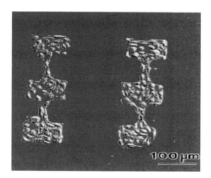

Fig. 2 Peptide ink. This type of peptide has three distinct segments: a functional segment where it interacts with other proteins and cells; a linker segment that is either flexible or stiff and sets the distance from the surface, and an anchor for covalent attachment to the surface. These peptides can be used as ink for an inkjet printer to directly print on a surface, instantly creating any arbitrary pattern, as shown here. Bovine aortic endothelial cells were confined to the patterns of squares connected with linear tracks. The patterns were made with an oxygen gas treated PDMS stamp to increase the surface hydrophilicity to facilitate EG6SH wetting

chor where a chemical group on the peptide can react with the surface to form a covalent bond [17].

Whitesides and coworkers developed a microcontact printing technology that combines semi-conducting industry fabrication, chemistry and polymer science to produce defined features on a surface down to the micrometer or nanometer scale [46–48]. Following microcontact printing, a surface can be functionalized with different molecules using a variety of methods which have now been modified with a variety of chemical compounds. Furthermore, peptides and proteins as inks have also been printed onto surfaces. This development has spurred new research into the control of molecular and cellular patterning, cell morphology and cellular interactions, and fueled new technology development. Peptide or protein inks have been directly printed on surfaces to allow adhesion molecules to interact with cells and adhere to the surface (Fig. 2) [49].

2.5
Peptide Surfactants/Detergents

Peptide surfactants or detergents stabilize membrane proteins, although membrane proteins make up approximately one-third of total cellular proteins and carry out some of the most important functions in cells, only several dozen membrane protein structures have been elucidated. This is in striking contrast to about 33 000 non-membrane protein structures that have been solved [50, 51].

The main reason for this delay is the difficulty in purifying and crystallizing membrane proteins because removal of lipids from membrane proteins

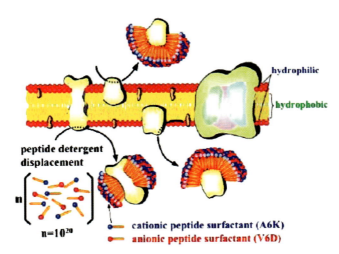

Fig. 3 Peptide surfactants of A6D and V6D. These simple self-assembling peptide surfactant/detergents can be used to solubilize, stabilize and crystallize membrane proteins

affects protein solubility and conformational stability. Although a variety of detergents and lipids as surfactants have been used to solubilize, stabilize and crystallize membrane proteins for several decades, these surfactants are still unable to significantly maintain structural stability of membrane proteins during experimental handling. In the other words, there is no "magic material" surfactant working on membrane proteins and there is an urgent need to develop new types of surfactants. We have used A6K (AAAAAAK) and V6D (VVVVVVD) to stabilize the photosynthetic protein-molecular complexes in solid-state devices and we showed that this new type of peptide detergents was very effective in stabilizing membrane protein functions, providing a powerful tool for membrane proteins research and application (Fig. 3) [10, 52–54].

2.6
Other Systems

Molecular self-assembly systems using nucleic acids on a chip have been developed. This new technology is based entirely on the principles of nucleic acid molecular self-assembly. Numerous new devices and technologies have been advanced. The most well-known example is the biochip technology "Lab on a Chip", "GeneChip", or "Microarray Technology" [55]. This microarray system is widely used in gene expression analysis, the human genome project, diagnostics, discovery of new functions of genes, and high-throughput drug discovery and screenings. In addition, people are now beginning to turn to testing the ability of peptide-based biomaterials to respond to external cues; this responsiveness has been collectively referred to as "smart behavior". Responsiveness can be defined at either the structural level or the functional level [56].

3
Fabrication of Nanomaterials Through Self-Assembling Systems

3.1
Nanofibers

The peptide Lego molecules can undergo self-assembly in aqueous solutions to form well-ordered nanofibers that further associate to form nanofiber scaffolds [15, 16, 57]. One of them, RADA16-I [58], is called PuraMatrix, because of its purity as a designed biological scaffold in contrast to other biologically derived scaffolds from animal collagen and Matrigel. Because these nanofiber scaffolds have 5–200 nm pores and have very high water content (99.5% or 5 mg/ml) (Fig. 4), they are useful in the preparation of 3D cell-culture media. The scaffolds closely mimic the porosity and gross structure

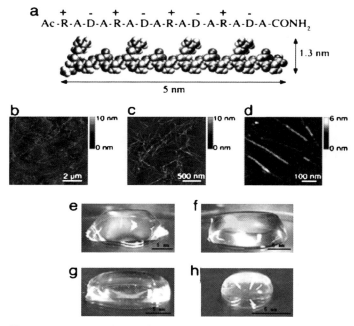

Fig. 4 Peptide RADA16-I. **a** Amino acid sequence and molecular model of RADA16-I. The dimensions are 5 nm long, 1.3 nm wide, and 0.8 nm thick; **b–d** AFM images of RADA16-I nanofiber scaffold. Note the different height of the nanofiber, $\approx$ 1.3 nm in **d**, suggesting a double-layer structure. **e–h** Photographs of RADA16-I hydrogel at various conditions: 0.5 wt % (pH 7.5) in **e**, 0.1 wt % (pH 7.5, Tris.HCl) in **f**, 0.1 wt % (pH 7.5, PBS) in **g** before sonication, and reassembled RADA16-I hydrogel after four rounds of sonication in **h**.

of extracellular matrices, allowing cells to reside and migrate in a 3D environment, and molecules, such as growth factors and nutrients, to diffuse in and out very slowly. These peptide scaffolds have been used for 3D cell culture, controlled cell differentiation, tissue engineering and regenerative medicine applications [59, 60].

3.1.1
Nanofibrils from α-helices

Several laboratories have designed fibrillar structures based on coiled-coil structural motifs, ranging from two-stranded to five-stranded coiled-coil structures [61–65]. In each case, investigators have recognized that peptides containing the coiled-coil motifs can self-assemble into a staggered interaction structure. Electrostatic interactions favor the formation of staggered arrangements of helices by two different 28-residue peptides. To help stabilize the staggered interactions, Woolfson's laboratory also took advantage of a buried asparagine residue in each of the two peptides, which can form structure stabilizing hydrogen bonds with its partner on the opposite strand only in the staggered conformation [61]. In addition, they have also described a clever synthetic method for introducing kinks and branches into fibrils [66, 67].

3.1.2
Nanofibrils from β-strands

We have studied sequences that form helical and sheet structures by incorporating specific interactions within a peptide sequence that would stabilize both sheet and helix formation [18]. In these sequences, such as DADADADARARARARA, a preformed β-sheet could be induced to adopt a α-helix in response to temperature and pH changes. Other groups have also studied this structural plasticity [45, 68]. In addition, investigators laid out a carefully reasoned strategy for the design of short hexapeptide sequences (i.e. KTVIIE, STVIIE, KTVIIT and KTVLIE) in order to test sequence elements critical for the formation of cross β-sheet structures and further test how polymeric β-sheets can mature into amyloid fibrils [69]. Towards this goal it is important to know how the cross β-sheet aggregates form and its role in neurodegenerative disease; recent efforts in the de novo design of peptide-based amyloid fibrils have aimed to identify simple sequences that minimally satisfy the requirements of fibril formation [69, 70].

3.2
Bionanotubes and Vesicles

These amphiphilic molecules readily interact with water and form various semi-enclosed environments. One of the best examples are phospholipids, the predominant constituents of the plasma membrane, which encapsulate and protect the cellular contents from the environment and are an absolute prerequisite for almost all living systems. Phospholipids readily undergo self-assembly in aqueous solution to form distinct structures that include micelles, vesicles and tubules. This is largely a result of the hydrophobic forces that drive the non-polar region of each molecule away from water and toward one another.

3.2.1
Short Amphiphilic Peptides

Our laboratory has designed a simple peptide system with those properties [52, 53]. We made short peptides of around six to seven amino acids that had the properties of surfactant molecules in that each monomer contained a polar and a non-polar region. For example, a peptide called A6D, the peptide molecule looked like a phospholipid in that it had a polar head group and a non-polar tail.

The homogeneity and size of the supramolecular assembly were sequence-sensitive: peptides of the same length behaved differently when they had different polar head or hydrophobic tail sequences. Such phenomena have been described theoretically and experimentally in other amphiphilic systems. The shape and size of the assemblies are ultimately dependent on the size and geometry of their constituents [71]. In order to visualize the structures in solution, we utilized the transmission electron microscope with the quick-freeze/deep-etch method for sample preparation [72], to preserve the structures that formed in solution for electron microscopy. We observed discrete nanotubes and vesicles

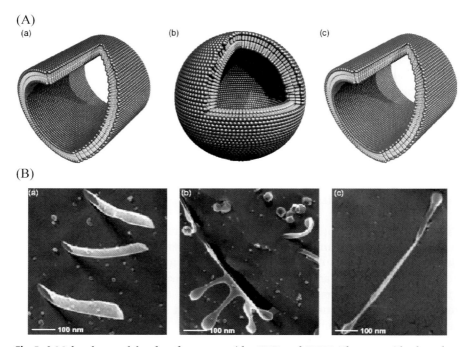

Fig. 5 A Molecular models of surfactant peptides V6D and K2V6. These peptides have hydrophilic heads; either negatively charged aspartic acid or positively charged lysine with hydrophobic valine tails [13–15]. **a** V6D in nanotube form. Billions of these molecules self-assemble to sequester the valine tails from water in **b** vesicle form or **c** nanotube form with positively charged heads. These nanostructures are rather dynamic undergoing assembly and disassembly. Color code: green-hydrophobic tails, red-aspartic acid, and blue-lysine. **B** Quick-freeze/deep-etch transmission electron micrographs of structures from surfactant peptides. **a** The nanotubes are clearly represented, with a diameter ~ 30–50 nm. **b** The nanotubes and vesicles are visible in the same frame suggesting that these structures are quite dynamic. It is plausible that the vesicles may be budded out from the nanotubes and/or they may fuse to form nanotubes in a reversible manner [13–15]. The diameter of these nanostructures is ~ 30–50 nm. **c** Phosphor-serine surfactant peptides form nano Q-tips

(Fig. 5) in the samples that gave homogeneous size distribution in the dynamic light scattering experiment. Those samples that were polydispersed tended to give irregular membranous layers. The nanotubes that formed had an average diameter of around 30 nanometers as examined by TEM, consistent with results obtained from the dynamic light scattering.

These nanotubes have the potential to act as templates for metallization and formation of nanowires. Furthermore, the nanovesicles may be useful as an encapsulating system for drug delivery. Chemical modification of the peptide monomer may expand the function of these structures. For example, a specific cell-surface ligand can be directly incorporated into a vesicle for targeted delivery of insoluble drugs to particular cells.

3.3
Nanometer-Thick

Molecular assembly can be targeted to alter the chemical and physical properties of a material's surface. Surface coatings instantly alter a material's texture, color, compatibility with and responsiveness to the environment. Conventional coatings are typically applied by painting or electroplating. Erosion is common mostly because the coatings are usually in the ten- and hundred micron size ranges and the interface is often not complementary at the molecular level [47, 73]. Peptides and proteins have also been printed onto surfaces which have now been modified with a vast family of chemical compounds; Mirkin and colleagues [74–76] have also developed dip-pen nanolithography to directly print micro- and nano-features onto surfaces. These developments have spurred new research into the control of molecular and cellular patterning, cell morphology, and cellular interactions [73, 77–79] and fueled new technology development.

Work in our laboratory has focused on designing a variety of peptides to self-assemble into a monolayer on surfaces and to allow adhesion molecules to interact with cells and adhere to the surface. Using proteins or peptides as ink, we have directly microprinted specific features onto the non-adhesive surface of polyethylene glycol to write any arbitrary patterns rapidly without preparing the mask or stamps (Fig. 6). This simple and rapid printing technology allowed us to design arbitrary patterns to address questions in neurobiology that would not have been possible before. Because understanding of correct complex neuronal connections is absolutely central to comprehension of our own consciousness, human beings are always interested in finding ways to further investigate this. However, the neuronal connections are exceedingly complex, and we must dissect the complex neuronal connections into smaller and more-manageable units to study them in a well-controlled manner through systematic biomedical engineering approaches. Therefore, nerve fiber guidance and connections can now be studied on special engineered pattern surfaces that are printed with protein and peptide materials [49].

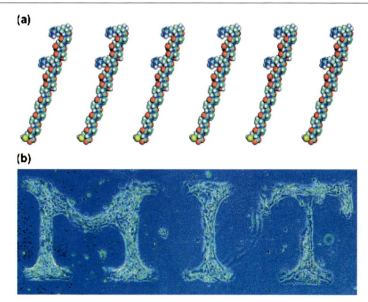

Fig. 6 Molecular structure of peptide ink. **a** This class of peptide ink has three general regions along their lengths: a ligand for specific cell recognition and attachment, a linker for physical separation from the surface, and an anchor for covalent attachment to the surface. Color code: carbon, *green*; hydrogen, *white*; oxygen, *red*; nitrogen, *blue*; thiol group, *yellow*. **b** Cells adhere to printed patterns. The protein was printed onto a uniform PEG inhibitory background. Cells adhered to patterns after 8–10 days in culture. (Images courtesy Sawyer Fuller and Neville Sanjana)

We are interested in studying the nerve fiber navigation on designed pattern surfaces in detail. Studies of nerve fiber navigation and nerve cell connections will undoubtedly enhance our general understanding of the fundamental aspects of neuronal activities in the human brain and brain–body connections. It will probably also have applications in screening neuropeptides and drugs that stimulate or inhibit nerve fiber navigation and nerve cell connections.

3.4
Nanowires

In the computing industry, the fabrication of nanowires using the "top-down" approach faces tremendous challenges. Thus, the possibility of fabricating conducting nanowires by molecular means using peptide scaffolds is of particular interest to the electronics industry. One can readily envision that nanotubes made from self-assembling peptides might serve as templates for metallization. Once the organic scaffold has been removed, a pure conducting wire is left behind and immobilized on a surface (Fig. 7). There is great inter-

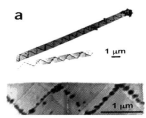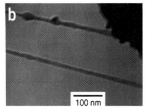

Fig. 7 Lipid, peptide and protein scaffold nanowires. **a** Lipid tubule–coated wire. Nanoparticles are coated on the left-handed helical lipid tubules. The nanoparticles are aligned inside the tubule along the regular helical pattern [66]. (Image courtesy of J. Schnur.) **b** Silver ions fill the nanotubes formed from a dipeptide, Phe-Phe, the shortest peptide possible. The silver alone formed a wire after removal of the Phe-Phe peptide scaffold [67]. (Image courtesy of M. Reches and E. Gazit.) **c** Discovery and selection of electronic materials using a bacteriophage display system. A combinatorial phage library was used to selectively bind to electronic materials. Selected recombinant phage peptide has a high affinity for GaAs. Fluorescently labeled phage has displaced the peptide specificity for GaAs and is capable of binding to the patterned GaAs nested in the square pattern on a wafer. The red line (1 μm in diameter) corresponds to GaAs and the black spaces (4 μm in diameter) are SiO$_2$. This peptide-specific binding could also potentially be used to deliver nanocrystals to specific locations [69]. (Image courtesy of A. Belcher.)

est in developing various methods for attaching conducting metal nanocrystals to a peptide for such a purpose.

3.4.1
From Nanotube to Nanowire

Matsui and colleagues [80] have reported success in functionalizing peptide nanotubes into nanowires. They not only coated the peptide nanotube with copper and nickel but also showed that their nanotubes can be coated with avidin, making them able to bind specifically to gold surfaces that have previously been treated with biotinylated self-assembled nanolayers. Lvov et al. [81] have fabricated nano- and microwires by coating the lipid tubules with silica and gold nanocrystals. They found that these nanocrystals are linked to the tubules according to the tubules' helical periodicity. These wires have been used for coating in a number of industrial applications. Reches and Gazit have demonstrated that a Phe-Phe dipeptide—the shortest peptide length possible, consisting of only two amino acids with a single amide bond—can form stable nanotubes. They then diffused silver ions into the defined tubes and were able to remove the peptide either enzymatically, chemically or through heat burning to reveal the silver wire [82].

In other recent work, amyloid protein nanofibers have been used as scaffold to align gold nanocrystals. Lindquist and colleagues [83] have reported how a bioengineered version of the prion-determining (NM) domain of the yeast prion protein Sup35 can provide a scaffold for fabricating nanowires, and tested the conducting capability of the resulting wires. These efforts collectively open a new direction in the fabrication of electronic nanomaterials.

3.4.2
Templates for Nanowires: DNA for Nano-Electronics

The DNA molecule has been suggested as a template for making nanoscale wires for the emergent field of nano-electronics. This is due to the regularity of the width of the DNA double helix and its robust mechanical properties. Several groups have succeeded in coating DNA molecules with metallic particles and have shown data on the conductive properties of these biotemplated materials. Braun et al. non-covalently bound a stretch (16 μm) of bacteriophage λ-DNA between two gold electrodes by allowing it to hybridize with short DNA fragments that had been covalently attached to those surfaces [84]. Electrical measurements indicated that the wires were non-conducting at low voltage bias, with resistances greater than the experimentally measurable 10^{13}. Furthermore, the shape of the I–V curve obtained was dependent on the voltage scan direction. Richter et al. employed a similar strategy to produce DNA-templated nanowires that showed relatively low resistances under low-voltage bias [85]. They reduced palladium on λ-DNA and immobilized the nanowire on gold electrodes. The resistances obtained were lower than 1 kv, with the specific conductivity approximately one order of magnitude lower than bulk palladium. Subsequently, the resistances of these palladium nanowires [86] were studied at low temperatures which discovered that the palladium metals reduced on a DNA template showed the expected quantum mechanical behavior, with their resistances increasing at low temperatures. This behavior is similar to that of thin palladium films and shows that wires templated with DNA molecules behave normally.

Mertig et al. discovered conditions in which fine and regular platinum clusters formed on DNA molecules by using first-principle molecular dynamics (FPMD), which ultimately yielded a faster rate of growth and finer metal clusters on the template [87]. Another metal that has been investigated for surface templating of DNA is gold. Harnack et al. investigated the binding and reduction of tris(hydroxymethyl) phosphine derivatized gold particles on calf-thymus DNA [88]. The rapidly formed nanowires show electrical conductivities about 1/1000th that of gold, which the authors attributed to the graininess of the material.

Patolsky et al. modified N-hydroxysuccinimide-gold nanoparticles with a nucleic-acid intercalating agent, amino psoralen [89]. In addition, Belcher and colleagues [75, 90–92] took a very different approach toward not only discovering, but also fabricating, electronic and magnetic materials, departing sharply from traditional materials process technology. Such approaches for producing finer and finer features at the nanoscale, with increasing density and in finite areas, may prove complementary to the microcontact printing process. Although the latter approach has become widely used and is rapidly being perfected, fabrication of the finest feature using microcontact printing is limited by the capabilities of lithography technologies currently used in the semiconductor industry.

3.5
Other Nanomaterials

The most advanced top-down technology for fabricating complex optical systems falls far short when compared with the accomplishments of living organisms at ambient temperature and low pressure (and without clean rooms) [93, 94]. Several groups have studied biomineralization in diverse marine organisms, notably the brittlestar Ophiocoma wendtii and the sponge Euplectella. Some remarkable living optical systems have been uncovered, such as the fiber-optical spicules from Euplectella that have the dimensions of a single human hair and can act as multimode waveguides. These discoveries have inspired Aizenberg et al. [95] to fabricate micropatterned single crystals and photonics with potential applications in communication technology. Christopher et al. report the synthesis and characterization of tethered PNA molecules (bisPNAs) designed to assemble two individual DNA molecules through Watson–Crick base pairing. The spacer regions linking the PNAs were varied in length and contained amino acids with different electrostatic properties [96, 97], their results indicate that the bisPNAs can be used for nanotechnology applications and that their favorable characteristics may lead to improved assemblies.

4
Application of Self-Assembling Systems

4.1
Simple Peptides Stabilize Mighty Membrane Proteins for Study

Cell membranes are largely made up of proteins, and membrane proteins account for about a third of all genes. Despite their importance, they are very hard to isolate and stabilize, which therefore prevents further understanding of membrane protein functions and related disease study. We have made a new type of peptide detergent and successfully stabilized the dauntingly large protein complex photosystem I (PS-I), an integral part of the photosynthetic machinery.

Two key technologies were employed to preserve the functionality of these photosynthetic complexes outside of their native environment. First, we added two peptide surfactants, one cationic A6K (AAAAAAK), and the other anionic V6D (VVVVVVD) into the photosynthetic complex fraction to stabilize it during device fabrication. Secondly, we examined the stability of PS-I by testing its fluorescence after attaching the detergent-protein complex to a glass slide [53, 54]. The intact PS-I emits red light with a characteristic peak wavelength as it degrades. This peak subsides and is replaced by another bluer peak. Even the two best standard surfactants did poorly at maintaining the red peak. In contrast, the spectrum after A6K extraction was almost a perfect match for the normal one, indicating the complex was largely intact after drying. Furthermore, the complex appeared to remain stable for up to three weeks on the glass slide. PS-I itself remains to be fully characterized, and this stabilization technique offers new means to explore its properties [98].

In addition, photosynthetic complexes are archetypal molecular electronic devices, containing molecular optical and electronic circuitry organized by a protein scaffold. Conventional technology cannot equal the density of the molecular circuitry found in photosynthetic complexes. Thus, if integrated with solid-state electronics, photosynthetic complexes might offer an attractive architecture for future generations of circuitry where molecular components are organized by a macromolecular scaffold. For utilization in practical technological devices they must be stabilized and integrated with solid-state electronics. Our results suggest that photosynthetic complexes may be used as an interfacial material in photovoltaic devices. Evolved within a thin membrane interface, photosynthetic complexes sustain large open circuit voltages of 1.1 V without significant electron-hole recombination, and they may be self-assembled into an insulating membrane, further reducing recombination losses. Peptide surfactants have been shown to stabilize these complexes during and after device fabrication. It is expected that the power conversion efficiency of a peptide-stabilized solid-state photosynthetic device may approach or exceed 20%. Similar integration techniques may apply to other biological or synthetic protein-molecular complexes [99].

These simply designed peptide detergents may now open a new avenue to overcome one of the biggest challenges in biology—to obtain large number of high resolution structures of membrane proteins. Study of the membrane proteins will not only enrich and deepen our knowledge of how cells communicate with their surroundings since all living systems respond to their environments, but these membrane proteins can also be used to fabricate the most advanced molecular devices, from energy harnessing devices to extremely sensitive sensors and medical detection devices.

4.2
Tissue Engineering

A new type of self-assembling peptide nano-fibril that serves as a substrate for neurite outgrowth and synapse formation is described (Fig. 8). The self-assembling peptide scaffolds are formed through the spontaneous assembly of ionic self-complementary β-sheet peptides under physiological conditions,

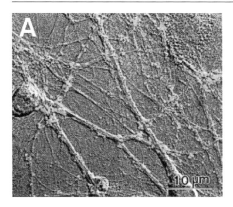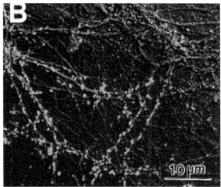

Fig. 8 Primary rat hippocampal neurons form active synapses on peptide scaffolds. The confocal images show bright discrete *green labeling* indicative of synaptically active membranes after FM1-43 incubation of neurons. **A** Active synapses on the peptide surface. **B** Active synapses on Matrigel. The active synapses on these different materials are not readily distinguishable, indicating that the peptide scaffold is a permissible substrate for synapse formation. *Bar* = 5–10 mm

producing a hydrogel material. The scaffolds can support neuronal cell attachment and differentiation as well as extensive neurite outgrowth. Furthermore, they are permissive substrates for functional synapse formation between the attached neurons. That primary rat neurons form active synapses on such scaffold surfaces in situ suggests these scaffolds could be useful for tissue engineering applications. The buoyant self-assembling peptide scaffolds with attached cells in culture can be transported readily from one environment to another. These biological materials created through molecular design and self-assembly may be developed as a biologically compatible scaffold for tissue repair and tissue engineering [100].

Self-assembling peptides are being developed as scaffolds for tissue regeneration purposes, including cartilage repair and the promotion of nerve cell growth [101]. A major benefit of synthetic materials is that they minimize the risk of biological contamination. Self-assembling peptides also frequently show favorable properties concerning biocompatibility, immunogenicity and biodegradability, producing non-toxic waste products. The amphiphilic peptide construct discussed above, containing a long hydrophobic tail linked to a cell-recognizing tag, can be customized for specific cell response by tailoring the sequence of the tag. Laminin is an extracellular matrix protein that influences neurite outgrowth. A peptide amphiphile shown to promote the re-growth of nerve cells in rats was made by including a neurite-promoting laminin epitope tag, IKVAV (C16-G3A4-IKVAV) [102]. Another construct, containing a heparin-binding site, shows very exciting preliminary results in being able to promote angiogenesis, the growth of blood vessels [103]. These types of peptide amphiphiles have been further modified with biotin [104] and a Gd^{3+} metal-chelating moiety suitable for detection by magnetic resonance imaging (MRI) [105].

Because an adequate blood supply to and within tissues is an essential factor for successful tissue regeneration, promoting a functional microvasculature is a crucial factor for biomaterials.

In 2005, Lee et al. demonstrated that short self-assembling peptides form scaffolds that provide an angiogenic environment promoting long-term cell survival and capillary-like network formation in three-dimensional cultures of human microvascular endothelial cells. Data showed that, in contrast to collagen type I, the peptide scaffold inhibited endothelial cell apoptosis in the absence of added angiogenic factors, accompanied by enhanced gene expression of the angiogenic factor VEGF. In addition, the results suggest that the process of capillary-like network formation and the size and spatial organization of cell networks may be controlled through manipulation of the scaffold properties, with a more rigid scaffold promoting extended structures with a larger inter-structure distance, as compared with more dense structures of smaller size observed in a more compliant scaffold. These findings indicate that self-assembling peptide scaffolds have potential for engineering vascularized tissues with control over angiogenic processes. Since these peptides can be modified in many ways, they may be uniquely valuable in the regeneration of vascularized tissues [106].

Emerging medical technologies for effective and lasting repair of articular cartilage include delivery of cells or cell-seeded scaffolds to a defect site to initiate de novo tissue regeneration. Biocompatible scaffolds assist in providing a template for cell distribution and extracellular matrix (ECM) accumulation in a three-dimensional geometry. A major challenge in choosing an appropriate scaffold for cartilage repair is the identification of a material that can simultaneously stimulate high rates of cell division and high rates of cell synthesis of phenotypically specific ECM macromolecules until repair evolves into steady-state tissue maintenance.

In 2002, we made a self-assembling peptide hydrogel scaffold for cartilage repair and developed a method to encapsulate chondrocytes within the peptide hydrogel. During 4 weeks of culture in vitro, chondrocytes seeded within the peptide hydrogel retained their morphology and developed a cartilage-like ECM rich in proteoglycans and type II collagen, indicative of a stable chondrocyte phenotype. Time-dependent accumulation of this ECM was paralleled by increases in material stiffness, indicative of deposition of mechanically functional neo-tissue. Taken together, these results demonstrate the potential of a self-assembling peptide hydrogel as a scaffold for the synthesis and accumulation of a true cartilage-like ECM within a three-dimensional cell culture for cartilage tissue repair.

In 2005, Lee and colleagues demonstrated that self-assembling peptides can be injected and that the resulting nanofiber microenvironments are readily detectable within the myocardium. Furthermore, the self-assembling peptide nanofiber microenvironments recruited progenitor cells that express endothelial markers, as determined by staining with isolectin and for the endothelial-specific protein platelet-endothelial cell adhesion molecule. Vascular smooth muscle cells were recruited to the microenvironment and appear to form functional vascular structures. After the endothelial cell population, cells that expressed sarcomeric actin and the transcription factor Nkx2.5 infiltrated the peptide microenvironment. When exogenous donor green fluorescent protein–positive neonatal cardiomyocytes were injected with the self-assembling peptides, transplanted cardiomyocytes in the peptide microenvironment survived and also augmented endogenous cell recruitment [107].

4.3
Gene and Drug Delivery

The lack of predictable safety and efficacy standards in somatic gene therapy systems, have brought the whole field to a crossroads. Replication-incompetent viruses, naked DNA injection and liposomal agents have been the predominant means of genetic transfer. To date, there has been little lasting impact in the typical practice of medicine conferred by these gene therapy technologies. The crux of today's gene therapy dilemma is still the same as it has always been: efficient, safe, targeted delivery and persistent gene expression [108, 109].

Peptide-based gene delivery agents are emerging as alternatives for safer in vivo delivery. The main attraction of these peptide systems is their versatility. Peptide-based delivery systems have the ability to deliver therapeutic proteins, bioactive peptides, small molecules and any size of nucleic acids. The use of these agents allows the researcher to intervene at multiple levels in the cells genetics and biochemistry and is a fundamental new technology in the gene therapy field [110, 111]. Peptide-delivery agents are more like traditional pharmacological drugs than gene therapy vectors. With the past to guide us, a critical re-evaluation of the best characteristics for an ideal delivery system is in order. The desirable features may include the items displayed in the paper [112].

We developed a series of surfactant peptides comprising a hydrophobic tail attached to a polar headgroup consisting of one to two positively charged residues at the C- or N-terminus, one example being LLLLLLKK. These peptides self-assemble in water to produce nanovesicles and nanotubes [54] as reported in a Science News commentary [113] and in recent reviews [114, 115]; these peptides have been used as DNA delivery vehicles. When placed in a solution of DNA, the positively charged peptides self-assembled into a nanotube or vessel, encapsulating the negatively charged DNA. This "minivan" was then able, at least in some cases, to deliver the DNA to growing cells as the minivan surface can be tagged with a marker that is specific to a particular cell type [113]. We expect that out of this emerging field, self-assembling peptide systems will play an increasing role in targeted molecular therapeutics and gene therapy.

4.4
Other Applications

Since peptides that can specifically bind to inorganic surfaces for particular applications that have no known analog in biology, molecular design may not be an efficient route to pursue. Even though one can potentially test many different biomolecular species to perform a particular function, the sheer number of samples that must be screened makes such an endeavor prohibitive in cost. For binding to GaAs (100), peptides with a higher number of uncharged polar and Lewis base side-chains became more predominant with successive rounds of selection. This could be attributed to the interaction of these functional groups with the Lewis acid sites of the GaAs surface.

Using a similar selection strategy, Lee et al. identified a bacteriophage that had the propensity to bind to ZnS crystal surfaces [116]. These phages were then mixed with ZnS quantum dots, forming a liquid crystalline suspension of the complex. This will push forward the areas of nano-electronic, optical and magnetic sciences and engineering.

Artificial peptide and protein libraries have been constructed for selection of novel proteins and peptide motifs that Nature never made [117–119]. Many investigators completely designed the peptide and protein libraries de novo, without a pre-existing protein basis. Although Nature has selected and evolved many diverse proteins for all sorts of functions that support life, it has not ventured into the functions outside of life. The protein universe is enormous, in comparison with what we know today. There are, undoubtedly, a great number of proteins that can exist beyond what has been founded in living systems. Numerous new proteins and peptides with desired and novel properties have been selected for a particular application. This strategy permits us to purposely select and rapidly evolve non-natural materials, nano-scaffolds and nano-construction motifs for a growing demand in nanotechnology. The numbers of these biologically based scaffolds are limitless and they will likely play an increasingly important role for the design of molecular machines, nanodevices and countless other novel, unanticipated new tools and applications.

5
Conclusions and Perspectives

From physics and engineering to biology, molecular design of self-assembling peptides is an enabling technology that will likely play an increasingly important role in the future of bionanotechnology and will change our lives in the coming decades. We have encountered many surprises since we started our serendipitous journey of working on various self-assembling peptide systems: from developing a class of pure peptide nanofiber scaffolds for cell engineering and for tissue repairing and tissue engineering, studying of the model system of protein conformational diseases, and designing peptide or protein inks for surface printing to finding peptide surfactants that solubilize and stabilize membrane proteins, bionanotubes and vesicles for delivering genes and drugs. Self-assembling peptide systems will create a new class of materials at the molecular scale and will have a high impact in many fields. We believe that application of these simple and versatile molecular self-assembly systems will provide us with new opportunities for studying complex and previously intractable biological phenomena.

References

1. Bränden CI, Tooze J (1999) Introduction to Protein Structure. Garland Publishing, New York
2. Petsko GA, Ringe D (2003) Protein Structure and Function. New Science Press Ltd, London, UK
3. Lehn JM (1995) Supramolecular Chemistry: Concepts and Perspectives. Wiley, New York
4. Seeman NC (2003) Nature 421:427–431
5. Seeman NC (2004) Sci Am 290:64–69
6. Petka WA, Harden JL, McGrath KP, Wirtz D, Tirrell DA (1998) Science 281:389–392
7. Nowak AP, Breedveld V, Pakstis L, Ozbas B, Pine DJ, Pochan D, Deming TJ (2002) Nature 417:424–428
8. Schneider JP, Pochan DJ, Ozbas B, Rajagopal K, Pakstis L, Kretsinger J (2002) J Am Chem Soc 124:15030–15037
9. Zhang S (2002) Biotech Adv 20:321–339
10. Zhang S (2003) Nat Biotechnol 21:1171–1178
11. Zhang S (2004) Nat Biotechnol 22:151–152
12. Zhang S, Altman M (1999) React Funct Polym 41:91–102
13. Zhang S, Rich A (1997) Proc Natl Acad Sci USA 94:23–28
14. Zhang S et al. (1992) EMBO J 11:3787–3796
15. Zhang S, Holmes T, Lockshin C, Rich A (1993) Proc Natl Acad Sci USA 90:3334–3338
16. Zhang S et al. (1995) Biomaterials 16:1385–1393
17. Zhang S et al. (1999) Biomaterials 20:1213–1220
18. Altman M, Lee P, Rich A, Zhang S (2000) Protein Sci 9:1095–1105
19. Zhang S, Altman M, Rich A (2001) In: Katzir E, Solomon B, Taraboulos A (eds) Diseases of Conformation—A Compendium. Bialik Institute, N. Ben-Zvi Printing Enterprises, Jerusalem, Israel, p 63–72
20. Zhang S, Marini D, Hwang W, Santoso S (2002) Curr Opin Chem Biol 6865–6871
21. Zhang S (2001) In: Buschow KH J, Cahn RW, Hemings MC, Ilschner B, Kramer EJ, Mahajan S (eds) Encyclopedia of Materials: Science and Technology. Elsevier, Oxford, p 5822

22. Hartgerink JD, Beniash E, Stupp SI (2001) Science 294:1684–1688
23. Ratner M, Ratner D (2003) Nanotechnology: A Gentle Introduction to the Next Big Idea. Prentice Hall, Upper Saddle River, New Jersey, USA
24. Gajdusek DC (1977) Science 197:943–960
25. Fandrich M, Fletcher MA, Dobson CM (2001) Nature 410:165–166
26. Scheibel T, Kowal AS, Bloom JD, Lindquist SL (2001) Curr Biol 11:366–369
27. Lynn DG, Meredith SC (2000) J Struct Biol 130:153–173
28. West MW, Wang W, Mancias JD, Beasley JR, Hecht MH (1999) Proc Natl Acad Sci USA 96:11211–11216
29. Hammarstrom P, Schneider F, Kelly JW (2001) Science 293:2459–2462
30. Shtilerman MD, Ding TT, Lansbury PT (2002) Biochemistry 41:3855–3860
31. Reches M, Porat Y, Gazit E (2002) J Biol Chem 277:35475–35480
32. Perutz MF, Pope BJ, Owen D, Wanker EE, Scherzinger E (2002) Proc Natl Acad Sci USA 99:5596–5600
33. Jimenez JL, Tennent G, Pepys M, Saibil HR (2001) J Mol Biol 311:241–247
34. Lindquist SL, Henikoff S (2002) Proc Natl Acad Sci USA 99(4):16377
35. Leon EJ, Verma N, Zhang SG, Lauåenburger DA, Kamm RD (1998) J Biomater Sci: Polym Ed 9:297–312
36. Selkoe DJ (1994) Annu Rev Neurosci 17:489–517
37. Roher AE, Baudry J, Chaney MO, Kuo YM, Stine WB, Emmerling MR (2000) Biochim, Biophys Acta 502:31–43
38. Wanker EE (2000) Mol Med Today 6:387–391
39. Žerovnik E (2002) Amyloid-fibril formation: Proposed mechanisms and relevance to conformational disease. Eur J Biochem 269:3362–3371
40. Luna-Muñoz J, Garíia-Sierrab F, Falcónc V, Menéndezc I, Chávez-Macías L, Mena R (2005) J Alzheimer's Disease 8:29–41
41. Hone E, Martinsa IJ, Jeoungd M, Ji TH, Gandy SE, Martins RN (2005) J Alzheimer's Disease 7:303–314
42. Hol WGJ, Halie LM, Sander C (1981) Nature 29:532–536
43. Minor DL, Kim PS (1996) Nature 380:730–734
44. Tan S, Richmond TJ (1998) Nature 391:660–666
45. Takahashi Y, Ueno A, Mihara H (1999) Bioorg Med Chem 7:177–185
46. Whitesides GM et al. (1991) Science 254:1312–1319
47. Mrksich M, Whitesides GM (1996) Annu Rev Biophys Biomol Struct 25:55–78
48. Chen CS et al. (1997) Science 276:1425–1428
49. Holmes TC et al. (2000) Proc Natl Acad Sci USA 97:6728–6733
50. Wallin E, von Heinje G (1998) Protein Sci 7:1029–1038
51. Loll PJ (2003) J Struct Biol 142:144–153
52. Vauthey S, Santoso S, Gong H, Watson N, Zhang S (2002) Proc Natl Acad Sci USA 99:5355–5360
53. Santoso S, Hwang W, Hartman H, Zhang SG (2002) Nano Lett 2:687–691
54. von Maltzahn G, Vauthey S, Santoso S, Zhang SG (2003) Langmuir 19:4332–4337
55. Orchid Biocomputers http://www.orchidbio.com
56. Rajagopal K, Schneider JP (2004) Curr Opin Struct Biol 14:480–486
57. Marini D et al. (2002) Nano Lett 2:295–299
58. Yokoi H, Takatoshi K, Zhang S (2005) Proc Natl Acad Sci USA 102:8414–8419
59. Kisiday J et al. (2002) Proc Natl Acad Sci USA 99:9996–10001
60. Goraman J (2000) Science News 158:23–24
61. Pandya MJ, Spooner GM, Sunde M, Thorpe JR, Rodger A, Woolfson DN (2000) Biochemistry 39:8728–8734

62. Potekhin SA, Melnik TN, Popov V, Lanina NF, Vazina AA, Rigler P, Verdini AS, Corradin G, Kajava AV (2001) Chem Biol 8:1025–1032
63. Zimenkov Y, Conticello VP, Guo L, Thiyagarajan P (2004) Tetrahedron 60:7237–7246
64. Wagner DE, Phillips CL, Lee LS, Ali WM, Nybakken EN, Crawford ED, Schwab AD, Smith WF, Fairman R (2005) Towards the development of peptide nanofilaments and nanoropes as smart materials. Proc Natl Acad Sci USA 102:12656–12661
65. Melnik TN, Villard V, Vasiliev V, Corradin G, Kajava AV, Potekhin SA (2003) Protein Eng 16:1125–1130
66. Ryadnov MG, Woolfson DN (2003) Angew Chem Int Ed 42:3021–3023
67. Ryadnov MG, Woolfson DN (2003) Nat Mater 2:329–332
68. Ciani B, Hutchinson EG, Sessions RB, Woolfson DN (2002) J Biol Chem 277:10150–10155
69. Lopez De La Paz M, Goldie K, Zurdo J, Lacroix E, Dobson CM, Hoenger A, Serrano L (2002) Proc Natl Acad Sci USA 99:16052–16057
70. Kammerer RA et al. (2004) Proc Natl Acad Sci USA 101:4435–4440
71. Israelachvili JN (1991) Intermolecular Surface Forces, 2nd edn. Academic Press, San Diego
72. Heuser J, Meth (1981) Cell Biol 22:97
73. Chen CS, Mrksich M, Huang S, Whitesides GM, Ingber DE (1997) Science 276:1425–1428
74. Piner RD, Zhu J, Xu F, Hong S, Mirkin CA (1999) Science 283:661–664
75. Lee K-B, Park SJ, Mirkin CA, Smith JC, Mrksich M (2002) Science 295:1702–1705
76. Demers LM, Ginger DS, Park S-J, Li Z, Chung S-W, Mirkin CA (2002) Science 296(5574):1836–1838
77. Mrksich M, Whitesides GM (1996) Annu Rev Biophys Biomol Struct 25:55–78
78. Dillo AK, Ochsenhirt SE, McCarthy JB, Fields GB, Tirrell M (2001) Biomaterials 22:1493–1505
79. Leufgen K, Mutter M, Vogel H, Szymczak W (2003) J Am Chem Soc 125:8911–8915
80. Djalali R, Chen YF, Matsui H (2002) J Am Chem Soc 124:13660–13661
81. Lvov YM, Price RR, Selinger JV, Singh A, Spector MS, Schnur JM (2000) Langmuir 16:5932–5935
82. Reches M, Gazit E (2003) Science 300:625–627
83. Scheibel T, Parthasarathy R, Sawicki G, Lin X-M, Jaeger H, Lindquist SL (2003) Proc Natl Acad Sci USA 100:4527–4532
84. Braun E, Eichen Y, Sivan U, Ben-Yoseph G (1998) Nature 391:775
85. Richter J, Mertig M, Pompe W, Monch I, Schackert HK (2001) Appl Phys Lett 78:536
86. Richter J, Mertig M, Pompe W, Vinzelberg H (2002) Appl Phys A Mater Sci Proc 74:725
87. Mertig M, Ciacchi LC, Seidel R, Pompe W, De Vita A (2002) Nano Lett 2:841
88. Harnack O, Ford WE, Yasuda A, Wessels JM (2002) Nano Lett 2:919
89. Patolsky F, Weizmann Y, Lioubashevski O, Willner I (2002) Angew Chem Int Ed 41:2323
90. Whaley SR, English DS, Hu EL, Barbara PF, Belcher AM (2000) Nature 405:665–668
91. Mao C, Flynn CE, Hayhurst A, Sweeney R, Qi J, Georgiou G, Iverson B, Belcher AM (2003) Proc Natl Acad Sci USA 100:6946–6951
92. Sarikaya M, Tamerler C, Jen AKY, Schulten K, Baneyx F (2003) Nat Materials 2:577–585
93. Aizenberg J, Tkachenko A, Weiner S, Addadi L, Hendler G (2001) Nature 412:819–822
94. Sundar VC, Yablon AD, Grazul JL, Ilan M, Aizenberg J (2003) Nature 424:899–900

95. Aizenberg J, Muller DA, Grazul JL, Hamann DR (2003) Science 299:1205–1208
96. Christopher J, Nulf, David RC (2002) Nucleic Acids Research 30:2782–2789
97. Winfree E, Liu F, Wenzler LA, Seeman NC (1998) Nature 394:539–544
98. Kiley P, Zhao XJ, Vaughn M, Baldo MA, Bruce BD, Zhang S (2005) PLoS BIOLOGY 3:1180–1186
99. Das R, Kiley PJ, Segal M, Norville J, Yu AA, Wang LY, Trammell SA, Reddick LE, Kumar R, Stellacci F, Lebedev N, Schnur J, Bruce BD, Zhang SG, Baldo M (2004) Nano Lett 4:1079–1083
100. Chen YC, Muhlrad A, Shteyer A, Vidson M, Bab I, Chorev M (2002) J Med Chem 45:1624–1632
101. Holmes TC (2002) Trends Biotechnol 20:16–21
102. Silva GA, Czeisler C, Niece KL, Beniash E, Harrington DA, Kessler JA (2004) Science 303:1352–1355
103. Service RF (2005) Science 308:44–45
104. Guler MO, Soukasene S, Hulvat JF, Stupp SI (2005) Nano Lett 5:249–252
105. Bull SR, Guler MO, Bras RE, Meade TJ, Stupp SI (2005) Nano Lett 5:1–4
106. Narmoneva DA, Oni O, Sieminski AL, Zhang S, Gertler JP, Kamm RD, Lee RT (2005) Biomaterials 26:4837–4846
107. Davis ME, Motion JPM, Narmoneva DA, Takahashi T, Hakuno D, Kamm RD, Zhang S, Lee RT (2005) Circulation 111:442–450
108. Anderson WF (1998) Nature 392:25–30
109. Gorecki DC, MacDermot KD (1997) Arch Immunol Ther Exp 45:375–381
110. Norman TC, Smith DL, Sorger PK, Drees BL, O'Rourke SM, Hughes TR, Roberts CJ, Friend SH, Fields S, Murray AW (1999) Science 285:591–595
111. Aramburu J, Yaffe MB, López-Rodríguez C, Cantley LC, Hogan PG, Rao A (1999) Science 285:2129–2133
112. Schwartz JJ, Zhang SG (2000) Curr Opin Mol Ther 2:162–167
113. Gorman J (2003) Sci News 163:43–44
114. Zhao X, Zhang S (2004) Trends Biotechnol 22:470–476
115. Zhang S, Zhao XJ (2004) J Mater Chem 14:2082–2086
116. Lee SW, Mao C, Flynn CE, Belcher AM (2002) Science 296:892
117. Blondelle SE, Houghten RA (1996) Trends Biotechnol 14:60
118. Moffet DA, Hecht MH (2001) Chem Rev 101:3191
119. Wei YT, Liu SL, Sazinsky DA, Moffet IP, Hecht MH (2003) Protein Sci 12:92

Adv Polym Sci (2006) 203: 171–190
DOI 10.1007/12_071
© Springer-Verlag Berlin Heidelberg 2006
Published online: 5 January 2006

Interfaces
to Control Cell-Biomaterial Adhesive Interactions

Andrés J. García

Woodruff School of Mechanical Engineering, Petit Institute for Bioengineering
and Bioscience, Georgia Institute of Technology, 315 Ferst Drive, IBB 2314,
Atlanta, GA 30332-0363, USA
andres.garcia@me.gatech.edu

Abstract Cell adhesion to adsorbed proteins and adhesive sequences engineered on surfaces is crucial to cellular and host responses to implanted devices, biological integration of biomaterials and tissue-engineered constructs, and the performance of biosensors, cell-based arrays, and biotechnological cell-culture supports. This review focuses on interfaces controlling cell-adhesive interactions, with particular emphasis on surfaces controlling protein adsorption, biomimetic substrates presenting bioadhesive motifs, and micropatterned surfaces to engineer adhesive areas. These approaches represent promising strategies to engineer cell-material biomolecular interactions in order to elicit specific cellular responses and enhance the biological performance of materials in biomedical and biotechnological applications.

Keywords Cell adhesion · Collagen · Fibronectin · Focal adhesions · Integrins

Abbreviations

COL-I   Type I collagen
ECM   Extracellular matrix
ELISA   Enzyme-linked immunosorbent assay
FN   Fibronectin
GFOGER   Glycine-phenylalanine-hydroxyproline-glycine-glutamate-arginine
LN   Laminin
PEG   Poly(ethelyne glycol)
RGD   Arginine-glycine-aspartic acid
SAM   Self-assembled monolayers
YIGSR   Tyrosine-isoleucine-glycine-serine-arginine

1
Cell Adhesion

1.1
Significance of Cell Adhesion

Cell adhesion to extracellular matrix (ECM) components is central to embryonic development, wound healing, and the organization, maintenance, and
repair of numerous tissues [1, 2]. Cell-matrix adhesive interactions provide
tissue structure and generate anchorage forces that mediate cell spreading
and migration, neurite extension, muscle-cell contraction, and cytokinesis [3–
5]. Moreover, cell adhesion triggers signals regulating the survival, cell-cycle
progression, and expression of differentiated phenotypes in multiple cell systems [2, 6, 7]. The critical importance of cell-ECM adhesion is underscored
by the absolute lethality at early embryonic stages in mice that have genetic
deletions for adhesion receptors, ligands, and adhesion-associated components [1, 8]. Furthermore, abnormalities in adhesive interactions are often
associated with pathological states, including blood-clotting and wound-
healing defects as well as malignant tumor formation [9, 10]. In addition to
pivotal roles in physiological and pathological processes, cell adhesion to adsorbed proteins or adhesive sequences engineered on surfaces is crucial to
cellular and host responses to implanted devices, biological integration of biomaterials and tissue-engineered constructs, and the performance of cell-based
arrays and sensors as well as biotechnological cell-culture supports [11–14].
Therefore, the development of biointerfaces that elicit specific cell-adhesive responses is central to numerous biomedical and biotechnological applications.

1.2
Integrin Adhesion Receptors

Integrins, a widely expressed family of glycosylated transmembrane receptors, constitute the primary adhesion mechanism to ECM components, in-

cluding fibronectin (FN), laminin (LN), and type I collagen (COL-I) [8]. In addition, several integrins bind to Ig-superfamily counterreceptors (e.g., VCAM, ICAM) to mediate cell–cell adhesion. Integrins are $\alpha\beta$ heterodimers; 18 α and 8 β subunits have been identified to dimerize into 24 distinct receptors. Most integrins are expressed on a wide variety of cell types, and most cells express several integrin receptors. However, some subclasses are only expressed in particular lineages, such as the leukocyte-specific β_2 integrins. The integrin receptor has a large extracellular domain formed by both α and β subunits, a single transmembrane pass, and two short cytoplasmic tails that do not contain catalytic motifs. The extracellular portions of the receptor also contain divalent metal-ion-binding sites, which are required for functional binding. Most integrins recognize short peptide sequences, such as the arginine-glycine-aspartic acid (RGD) motif present in many ECM pro-

Table 1 Selected integrins and their ligands

Integrin	Ligand	Binding site
$\alpha_1\beta_1$	COL-IV	CNBr frag. a1(IV)2
	LN	E1-4, P1
$\alpha_2\beta_1$	COL-I	GFOGER (triple helix)
$\alpha_3\beta_1$	LN	E3, GD6 peptide
	Thrombospondin	TSP-768
$\alpha_4\beta_1$	FN	IIICS (EILDV, REDV)
	Osteopontin	Hep II (IDAPS)
$\alpha_5\beta_1$	FN	RGD + PHRSN
$\alpha_6\beta_1$	LN	E8
$\alpha_{IIb}\beta_3$	Fibrinogen	RGD (a); KQAGD (g)
	FN	RGD
	Vitronectin	RGD
	von Willebrand factor	RGD
$\alpha_V\beta_3$	FN	RGD
	Vitronectin	RGD
	Fibrinogen	RGD
	von Willebrand factor	RGD
	Thrombospondin	RGD
	Osteopontin	RGD
	Bone sialoprotein	RGD
	Tenascin	RGD
	COL (nonfibrillar)	RGD
$\alpha_M\beta_2$	Fibrinogen	P1, P2
	iC3b	
	Factor X	

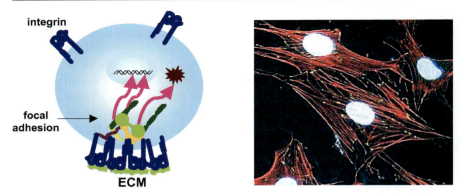

Fig. 1 Cell adhesion to ECM components involves binding of integrin receptors. Following binding, integrins cluster, interact with the actin cytoskeleton, and form focal adhesions, supramolecular complexes containing structural and signaling components. Signals from focal adhesions regulate protein activity and gene expression (diagram, *left*). Immunofluorescence staining (*right*) for cells spreading on FN (*blue* – DNA; *red* – F-actin cytoskeleton; *green* – vinculin)

teins including FN and vitronectin, and these motifs often contain an acidic amino acid. Ligand specificity is dictated by both subunits of a given $\alpha\beta$ heterodimer, and in many instances individual integrins can bind to more than one ligand (Table 1).

Integrin-mediated adhesion is a highly regulated process that involves receptor activation and mechanical coupling to extracellular ligands [4, 15, 16]. Integrins undergo conformational changes between high-affinity ("ON") and low-affinity ("OFF") states that provide for spatial and temporal control of ligand binding activity. Following activation, bound receptors rapidly associate with the actin cytoskeleton and cluster together to form focal adhesions, discrete supramolecular complexes that contain structural proteins, such as vinculin, talin, and α-actinin, and signaling molecules, including FAK, Src, and paxillin (Fig. 1) [17]. Interestingly, there are differences in the state of activation and components of focal adhesive structures, possibly reflecting different functional complexes [18]. Focal adhesions are central elements in the adhesion process, functioning as structural links between the cytoskeleton and ECM to generate mechanical forces mediating stable adhesion, spreading, and migration. Furthermore, in combination with growth factor receptors, focal adhesions activate signaling pathways, such as MAPK and JNK, that regulate transcription factor activity and direct cell cycle progression and differentiation [6]. For example, binding of integrins $\alpha_5\beta_1$ to FN and $\alpha_2\beta_1$ to COL-I directs osteoblast cell survivial, proliferation, bone-specific gene expression, and matrix mineralization [19–21].

1.3
Adhesive Interactions in Cell and Host Responses to Biomaterials

Because of their essential roles in cell adhesion to ECM components, integrins are critically involved in host and cellular responses to biomaterials. For example, the platelet integrin $\alpha_{IIb}\beta_3$ (GP IIb/IIIa) binds to several ligands involved in platelet aggregation in hemostasis and thrombosis, such as fibrinogen, von Willebrand factor, and fibronectin [8]. Furthermore, this receptor mediates initial events in the blood-activation cascade upon blood contact with synthetic materials [22, 23]. Leukocyte-specific β_2 integrins, in particular $\alpha_M\beta_2$ (Mac-1), mediate monocyte and macrophage adhesion to various ligands, including fibrinogen, fibronectin, IgG, and complement fragment iC3b, and these receptors play central roles in inflammatory responses in vivo [24, 25]. Binding of $\alpha_M\beta_2$ integrin to fibrinogen P1 and P2 domains exposed upon adsorption to biomaterial surfaces controls recruitment and accumulation of inflammatory cells on implanted devices [26]. This integrin is also involved in macrophage adhesion and fusion into giant foreign-body cells [25, 26]. For numerous connective, muscular, neural, and epithelial cell types, β_1 integrins provide the dominant adhesion mechanism to extracellular matrix ligands, including proteins adsorbed onto biomaterial surfaces [27]. In addition to supporting adhesion, spreading, and migration, these receptors activate intracellular signaling pathways controlling gene expression and protein activity that regulate cell proliferation and the expression of differentiated phenotypes.

Integrins mediate cellular interactions with biomaterials by binding to adhesive extracellular ligands that can be (i) adsorbed from solution (e.g., protein adsorption from blood, plasma, or serum); (ii) secreted and deposited onto the biomaterial surface by cells (for example, FN and COL-I deposition); and/or (iii) engineered at the interface (e.g., bioadhesive motifs such as RGD incorporated onto synthetic supports) (Fig. 2). These interactions are often highly dynamic in nature, and the dominant adhesion mechanism may change over time and for different cell types. For example, the dominant adhesive ligand present on biomaterials when exposed to plasma is fibrinogen, while vitronectin is generally responsible for cell adhesion to surfaces exposed to serum [28, 29]. These adhesive ligands may be displaced and replaced by other adhesive proteins in the surrounding medium. Additionally, while cells may initially adhere to synthetic surfaces via proteins precoated (e.g., FN treatment) or adsorbed from solution, many cell types rapidly degrade/reorganize this layer of adsorbed proteins and deposit their own ECM. Furthermore, the integrin expression and activity profiles on a particular cell can change over time. As mentioned above, most cells exhibit several integrins specific for the same ligand, and the binding activity of these receptors can be rapidly regulated via changes in integrin conformation. It is important to note that the integrin expression profile does not necessarily correlate

Fig. 2 Integrins mediate cellular interactions with biomaterials by binding to adhesive extracellular ligands that can be (i) adsorbed from solution, (ii) deposited onto the surface by cells, and/or (iii) engineered at the interface (e.g., bioadhesive motifs such as RGD incorporated onto synthetic supports). Adapted from [53]

with integrin function on a particular substrate. Finally, multiple integrins are typically involved in a particular cellular response. For example, initial monocyte adhesion to biomaterials is mediated primarily by β_2 integrin, while both β_1 and β_2 integrins are involved in macrophage adhesion and fusion into foreign-body giant cells [30].

2
Surfaces Controlling Protein Adsorption and Activity

The chemical and topographical characteristics of surfaces have profound effects on cellular, tissue, and host responses to synthetic materials [11, 31]. Consequently, surface modifications of chemistry and roughness have been introduced to improve performance in virtually all materials used in biotechnological [e.g., tissue culture and enzyme-linked immunosorbent assay (ELISA) plates, gene and protein array chips, bioseparation and bioprocess matrices] and biomedical (e.g., vascular grafts, orthopedic and dental implants, biosensors, catheters) applications. This review focuses on interfaces controlling cell-biomaterial adhesive interactions via manipulations of material surface chemistry to modulate protein adsorption and activity.

2.1
Protein Adsorption in Cell-Biomaterial Interactions

Protein adsorption onto synthetic surfaces plays central roles in numerous biomedical and biotechnological applications. Adsorption of blood components onto material surfaces triggers coagulation and complent activation as well as providing adhesive ligands mediating inflammatory responses to implanted devices. As discussed previously, cell adhesion to synthetic surfaces, including tissue-culture supports, tissue-engineering scaffolds, and affinity chromatography media, often involves binding of cellular receptors to proteins adsorbed onto the biomaterial support. In addition, protein adsorption considerations are critical to various classes of biosensors, where nonspecific adsorption (fouling) typically limits sensor performance. Hence, adsorbed proteins function as signal transduction elements at the interface of the material and the biological system.

Protein adsorption is a complex, dynamic, energy-driven process involving noncovalent interactions, including hydrophobic interactions, electrostatic interactions, hydrogen bonding, and van der Waals forces [32, 33]. Protein parameters such as primary structure, size, and structural stability and surface properties including surface energy, roughness, and chemistry have been identified as key factors influencing the adsorption process. Furthermore, multicomponent systems, such as plasma and serum, exhibit dynamic adsorption profiles. In this phenomenon, known as the Vroman effect, the protein film at the interface changes over time as proteins in high concentration adsorb first but are subsequently displaced by proteins that have higher affinitiy for the surface [32]. Therefore, adsorption from protein mixtures is selective and leads to enrichment of the surface phase in particular proteins. In addition to differences in adsorbed density, many proteins undergo changes in structure upon adsorption, and these structural changes alter their biological activity. Thus, analyses of protein adsorption must consider adsorbed protein species (for multicomponent systems), density, and biological activity. Finally, while most detailed studies of protein adsorption continue to be experimental in nature, new computational approaches are expected to provide insights into mechanisms controlling protein adsorption at the molecular level [34–36].

2.2
Surfaces That Resist Protein Adsorption

The generation of nonfouling surfaces that resist the nonspecific adsorption of biomolecules is critical to the biological performance of numerous biomedical devices, including blood-contacting devices, catheters, and sensing/stimulating leads [33]. In addition, nonfouling surfaces are important to in vitro applications such as oligonucleotide, protein, and cell arrays. The motivation for the development of these nonfouling surfaces is that prevention of protein adsorption will minimize cell adhesion and inflammatory responses and result in improved device performance. Despite considerable research efforts over the last three decades, robust surface treatments that *completely* eliminate protein adsorption over the lifetime of a device have not been obtained. Nevertheless, significant progress has been attained in understanding the mechanisms driving protein adsorption, and several chemical groups that resist protein adsorption have been identified. A key element in resistance to protein adsorption is the energetics of interfacial solvent water molecules, i.e., hydration layers associated with the proteins and the surface. For example, it is generally agreed that the major driving force for the irreversible adsorption of proteins onto hydrophobic surfaces is the unfolding of the protein and subsequent release of "bound" water molecules, which provides a huge increase in the entropy of the system favoring protein adsorption. Therefore, surfaces that retain interfacial water molecules, i.e., present an interface that "looks like" bulk water, should have low protein adsorption. Based on this inference, most common approaches to reducing protein adsorption onto biomaterial surfaces involve treatments that render surfaces more hydrophilic. In fact, simple treatments with hydrophilic biomolecules, such as albumin, casein, dextran, and even lipid bilayers, generally reduce protein adsorption to low levels. However, these treatments lose their nonfouling properties over time due to displacement by other proteins and lipids and/or cell-mediated degradation.

Poly(ethelyne glycol) (PEG) ($- [CH_2CH_2O]_n$) groups have proven to be the most protein-resistant functionality and remain the standard for comparison [37]. A strong correlation exists between PEG chain density and length and resistance to protein adsorption, and consequently cell adhesion [38, 39]. The mechanism of resistance to protein adsorption of PEG surfaces probably involves a combination of the ability of the polymer chain to retain interfacial water ("osmotic repulsion") and the resistance of the polymer coil to compression due to its tendency to remain as a random coil ("entropic repulsion") [33]. Well-packed, self-assembled monolayers (SAMs) of EG repeats as short as three repeats display excellent nonfouling characteristics [40, 41]. The nonfouling properties of these surfaces are dependent on the conformation of the oligoEG chain—a helical or amorphous conformation exhibits significantly higher resistance to protein adsorption compared to an all trans conformation, probably due to stronger EG-interfacial water interactions [42]. Other hydrophilic polymers, such as poly(2-hydroxyethyl methacrylate), polyacrylamide, and phosphoryl choline polymers, also resist protein adsorption [33]. In addition, mannitol, oligomaltose, and taurine groups have emerged as promising moieties to prevent protein adsorption [43–45]. Nevertheless, more comprehensive analyses, including in vivo studies, are required to establish the efficacy and applicability of these approaches in preventing protein adsorption and biofouling.

2.3
Substrates Modulating Adsorbed Protein Activity

Surface modifications to enhance protein adsorption and cell adhesion have been extensively pursued to improve device performance for both in vitro and in vivo applications. Everyday examples are tissue-culture-treated polystyrene and substrates for enzyme-linked immunosorbent assays (ELISA). In these applications, the base polymer is treated to reduce hydrophobicity and improve cell adhesion, as for tissue-culture-treated substrates, or modified to enhance protein adsorption in order to increase signal detection by antibodies in ELISA plates.

A promising strategy to direct cellular responses is to engineer surfaces that control the biological activity of adsorbed proteins. Using SAMs of ω-functionalized alkanethiols on gold to present well-defined chemistries (CH$_3$, OH, COOH, NH$_2$), García and colleagues demonstrated that surface chemistry modulates the structure of adsorbed FN [46]. The structure of the cell-binding domain of FN, which includes the integrin-binding RGD site, is particularly sensitive to the underlying support chemistry. These surface-dependent differences in FN structure alter integrin receptor binding, resulting in selective binding of $\alpha_5\beta_1$ integrin on OH and NH$_2$ surfaces, binding of both $\alpha_5\beta_1$ and $\alpha_V\beta_3$ in the COOH surface, and poor binding of either integrin on the CH$_3$ support [46] (Fig. 3). Surface-chemistry-dependent differences in integrin binding differentially regulate focal adhesion assembly in terms of molecular composition and signaling [47]. Furthermore, differences in integrin binding specificity modulate osteoblastic differentiation and mineralization [48] (Fig. 3). Biomaterial-chemistry-dependent differences in integrin binding specificity also regulate the switch between myogenic proliferation and differentiation [49], demonstrating a general surface engineering approach to control cell function. This strategy of biomaterial-directed con-

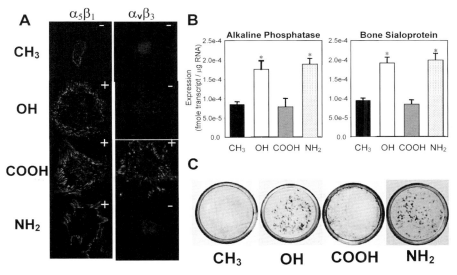

Fig. 3 Biomaterial surface chemistry modulates cellular responses. **A** SAMs presenting different chemistries differentially modulate integrin receptor binding in osteoblasts. **B** Substrate-dependent differences in osteoblast-specific gene expression correlate with integrin binding specificity. **C** Matrix mineralization is dependent on integrin binding specificity. Surfaces that support specific binding of $\alpha_5\beta_1$ integrin exhibit high levels of mineralization. Adapted from [46, 48]

trol of integrin binding specificity could be exploited to precisely engineer cell-material biomolecular interactions to activate specific signaling pathways and differentiation programs.

3
Biomimetic Interfaces Promoting Cell Adhesion

3.1
Biological Motifs as Targets for Biomaterial Applications

Significant advances in the engineering of biomaterials that elicit specific cellular responses have been attained over the last decade by exploiting biomolecular recognition. These biomimetic engineering approaches focus on integrating recognition and structural motifs from biological macromolecules with synthetic and natural substrates to generate materials with biofunctionality [14, 50]. These strategies represent a paradigm shift in biomaterials development from conventional approaches dealing with purely synthetic or natural materials to hybrid materials incorporating biological motifs. These biomimetic strategies provide promising schemes for the development of novel bioactive substrates for enhanced tissue replacement and regeneration. Because of the central roles that ECMs play in tissue morphogenesis, homeostasis, and repair, these natural scaffolds provide several attractive characteristics worthy of copying or mimicking to convey functionality for molecular control of cell function, tissue structure, and regeneration. Four ECM "themes" have been targeted: (i) motifs to promote cell adhesion, (ii) growth factor binding sites that control presentation and delivery, (iii) protease-sensitive sequences for controlled degradation, and (iv) structural motifs to convey mechanical properties. This review focuses on bioadhesive materials; excellent reviews on other biomimetic strategies can be found elsewhere [14].

3.2
First-Generation Biomimetic Adhesive Supports: Short Oligopeptides

Following the identification of adhesion motifs from ECM components, such as the RGD sequence in FN and the tyrosine-isoleucine-glycine-serine-arginine (YIGSR) oligopeptide in LN, short bioadhesive oligopeptides have been tethered/immobilized onto synthetic or natural substrates and three-dimensional scaffolds to produce biofunctional materials that bind integrin receptors and promote adhesion in various cell types [51–53] (Fig. 4). Non-fouling supports, such as PEG, polyacrylamide, and alginate, are often used to reduce nonspecific protein adsorption and present the bioadhesive motif within a nonadhesive background. Tethering of these short bioactive sequences promotes in vitro cellular activities, including adhesion, migration,

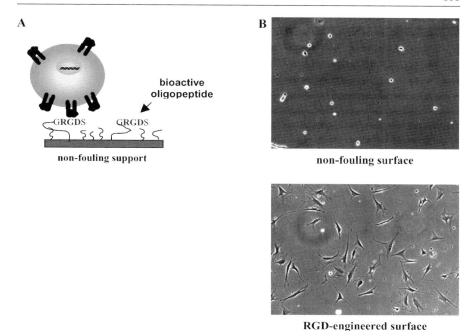

Fig. 4 Tethering of short bioadhesive peptides onto nonfouling surfaces supports cell-adhesive activities. **A** Schematic diagram showing specific integrin binding to bioadhesive RGD motif. **B** RGD immobilization onto nonfouling support promotes cell adhesion and spreading

and expression of differentiated phenotypes in multiple cellular systems. The density of tethered peptides is an important design parameter as cell adhesion, focal adhesion assembly, spreading and migration, neurite extension, and cell differentiation exhibit peptide-density-dependent effects [54–61]. More importantly, these biomimetic approaches enhance tissue regeneration in vivo, including as bone and cartilage formation, peripheral-nerve regeneration, and corneal tissue repair [62–67].

The use of short oligopeptides derived from ECM biomolecules presents advantages over the native biomolecules, such as conveying biospecificity while avoiding unwanted interactions with other regions of the native ligand, facile incorporation into synthetic and natural backbones under conditions incompatible with most biomacromolecules, and enhanced stability. The early successes with biomaterials displaying short bioadhesive oligopeptides established the potential of this biomolecular engineering strategy as a route to generate biointerfaces that interact with cells in prescribed and specific fashions. Nonetheless, functionalization of biomaterials with short bioadhesive motifs is limited by (i) reduced activity of oligopeptides compared to native biomacromolecule due to the absence of complementary or modulatory domains, (ii) limited specificity among integrin adhesion receptors, and (iii) inability to bind certain receptors due to conformational differences compared to the native ligand. These limitations are critical shortcomings because specific integrin receptors trigger different signaling pathways and cellular programs [48, 68–72]. Consequently, "second-generation" bioligands have been pursued to address the limitations associated with short bioadhesive oligopeptides.

3.3
Second-Generation Biomimetic Adhesive Supports: Ligands with Integrin Specificity

Engineered ligands, both short oligopeptides and recombinant protein fragments, incorporating additional residues or/and structural characteristics mimicking the native ligand have been developed to convey receptor specificity among RGD-binding integrins (Fig. 5). As discussed in Sect. 2.3, binding of *specific* integrin receptors can be exploited to regulate distinct cellular outcomes. Inclusion of flanking residues and constraining the conformation of the RGD motif to a loop via cyclization improve ligand specificity for integrins [73–75]. Nevertheless, these short peptides are limited in their ability to support specific integrin binding. For example, RGD domains in a loop conformation similar to FN bind $\alpha_V \beta_3$ but support poor $\alpha_5 \beta_1$ binding when compared to native FN [76]. Binding of $\alpha_5 \beta_1$ requires both the PHSRN sequence in the 9th type III repeat and RGD motif in the 10th type III repeat of FN [77]. Each domain independently contributes little to binding, but in combination, they synergistically bind to $\alpha_5 \beta_1$ [78, 79]. In efforts to include this essential PHSRN synergy site outside the RGD binding motif in fibronectin, mixtures of RGD and PHSRN peptides, either independently or within the same backbone, have been tethered onto nonfouling supports [80, 81]. Although these ligands support integrin binding and cell adhesion, their activity has not been directly compared to FN. Due to the high sensitivity of $\alpha_5 \beta_1$-FN binding to small perturbations in the structural alignment of these domains [70, 82], reconstitution of the proper binding structure using short peptides remains a challenging task. As an alternative to these synthetic routes, recombinant FN fragments spanning the 9th and 10th type III repeats have been tethered onto supports or incorporated into peptide backbones [83, 84]. These engineered ligands support robust $\alpha_5 \beta_1$-mediated adhesion and focal adhesion assembly at levels comparable to native FN (Fig. 5). In addition to providing increased specificity over linear RGD peptides, the use of recombinant fibronectin fragments offers several advantages compared to whole FN, including reduced antigenicity, elimination of domains that may elicit undesirable reactions, and enhanced cost efficiency. Recombinant fragments also provide flexibility in the engineering of specific characteristics on the fragment via site-directed mutagenesis in order to enhance tethering and activity.

A

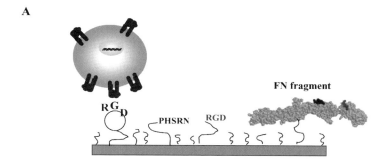

B

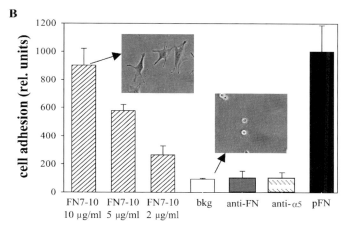

Fig. 5   Second-generation biomimetic adhesive supports. **A** Schematic showing major strategies pursued to improve integrin binding specificity. **B** A recombinant fragment of FN (FN7-10) containing the PHSRN and RGD binding sites supports dose-dependent levels of $\alpha_5\beta_1$ integrin-mediated adhesion. Adhesion levels are comparable to the native ligand plasma FN (pFN) and are completely blocked by antibodies against the binding site in FN (anti-FN) or $\alpha_5\beta_1$ integrin (anti-α_5). Adapted from [53, 83]

Non-RGD binding integrins are also critical to many cellular activities and, thus, represent important targets for therapeutic manipulations. For example, the collagen-binding integrin $\alpha_2\beta_1$ regulates various cellular activities, including adhesion, migration, proliferation, and differentiation in osteoblasts, keratinocytes, smooth muscle cells, and platelets [85]. Integrin $\alpha_2\beta_1$ recognizes the glycine-phenylalanine-hydroxyproline-glycine-glutamate-arginine (GFOGER) motif in residues 502–507 of the $\alpha_1[I]$ chain of COL-I [86]. Integrin recognition is entirely dependent on the triple-helical conformation of the ligand similar to that of native collagen. Tethering of a triple helical peptide incorporating the GFOGER motif to surfaces promotes $\alpha_2\beta_1$-mediated adhesion, focal adhesion signaling, and osteoblast differentiation to levels comparable to COL-I-coated supports [87, 88]. These results indicate that integrin binding specificity can be conveyed by engineering ligands that recapitulate the secondary and tertiary structure of the natural biopolymers (Fig. 6). The improved activity/selectivity of these "second-generation" biomolecular interfaces enhances the therapeutic and biotechnological potential of biomimetic materials.

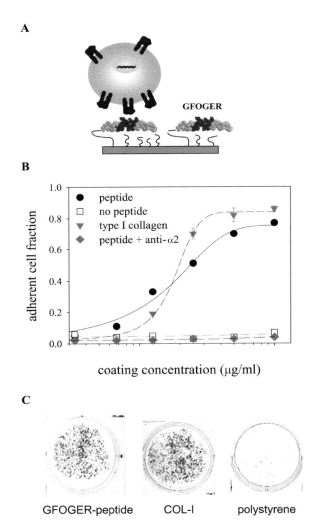

Fig. 6 Ligands with secondary/tertiary structure promote binding of $\alpha_2\beta_1$ integrin, a non-RGD binding integrin. **A** Diagram showing strategy for presenting collagen-mimetic, triple-helical GFOGER peptide. **B** Tethering of GFOGER onto nonfouling surfaces supports cell adhesion comparable to COL-I, and adhesion is completely blocked by antibodies against $\alpha_2\beta_1$ integrin. **C** Equivalent levels of matrix mineralization for osteoblasts grown on GFOGER-functionalized and COL-I-coated surfaces. Adapted from [87, 88]

4
Micropatterned Supports to Control Cell Adhesion

4.1
Engineering Cell Shape and Adhesive Area

Micropatterning techniques have been extensively applied to engineer cell position, shape, and adhesive area [89, 90]. These approaches generally rely on creating domains that readily adsorb proteins, and hence are cell-adhesive, and are surrounded by a nonfouling, nonadhesive background. These micropatterned supports can be easily generated by conventional photolithography as well as "soft" lithography approaches, including microcontact printing. In addition, direct protein stamping has been applied to create cell adhesive domains, but the stability of these patterns is limited by cell-mediated ECM reorganization and deposition. These substrates with defined adhesive areas have been exploited to analyze the roles of cell shape and cell–cell interactions on cell survival, expression of tissue-specific markers, and commitment to differentiated lineages [91–94]. Conversely, micropatterned substrata allow engineering of adhesive area, and in particular focal adhe-

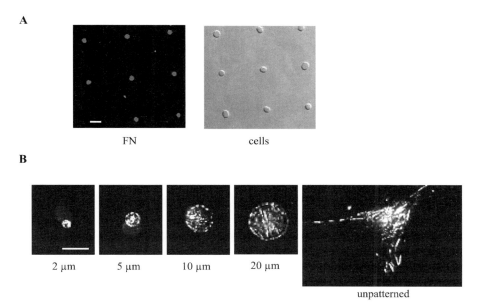

Fig. 7 Micropatterned surfaces to engineer focal adhesion size. **A** Adhesive islands within nonfouling background showing preferential FN adsorption and cell adhesion. Cells adhere and remain constrained to micropatterned island. *Bar*: 20 μm. **B** Vinculin localization to micropatterned domain, showing precise control over focal adhesion assembly. *Bar*: 10 μm. Adapted from [95, 99]

sion size, while maintaining a constant cell shape constant in order to analyze the contributions of cell-substrate contact area to adhesive processes such as adhesion strength and spreading [95, 96] (Fig. 7). Finally, micropatterning approaches provide robust tools for the creation of cellular arrays for high-throughput screening [97, 98].

4.2
Adhesion Strengthening Responses to Micropatterned Surfaces

Functional analyses of cell adhesion strengthening on micropatterned substrates provide an excellent illustration of the ability to engineer cell-material interactions via surface engineering. Previous analyses of cell adhesion strengthening have been limited by time-dependent changes in adhesive area,

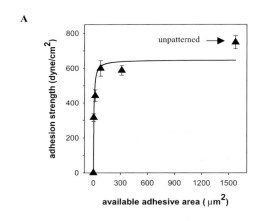

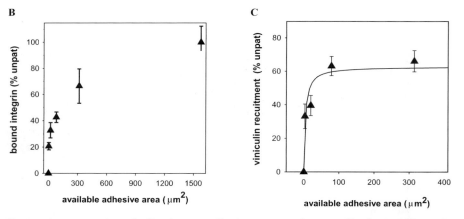

Fig. 8 Micropatterning of cell-substrate adhesive area regulates **A** cell adhesion strength, **B** integrin binding, and **C** focal adhesion assembly. Adapted from [99]

cell shape, and focal adhesion assembly. In a recent study, microcontact printing of SAMs was used to generate arrays of circular adhesive islands surrounded by a nonadhesive background to analyze the role of adhesive area on adhesion strengthening [99]. The use of micropatterned surfaces affords precise control over adhesive area, cell spreading/shape, and the position and size of focal adhesions, allowing decoupling of cell shape/spreading from focal adhesion formation. Cells individually adhere to the adhesive islands and maintain a nearly spherical shape, while the cell-substrate adhesive area conforms to the pattern dimensions (Fig. 7). Adhesion strength exhibits hyperbolic increases with available contact area, reaching a saturation value equivalent to the strength of unpatterned cells (Fig. 8). Moreover, integrin binding and focal adhesion assembly on the engineered adhesive ligand display nonlinear increases with available contact area, approaching saturating levels at high adhesive areas (Fig. 8). These results demonstrate precise control over adhesive interactions in terms of molecular events (integrin binding and focal adhesion assembly) and functional outcomes (adhesion strength).

5
Conclusions and Future Prospects

Surface-engineering approaches focusing on controlling cell-adhesive interactions represent promising strategies to engineer cell-biomaterial biomolecular interactions in order to elicit specific cellular responses and enhance the biological performance of materials in biomedical and biotechnological applications. While considerable progress has been made in developing surfaces that control protein adsorption and substrates that present biomimetic motifs, next-generation bioadhesive interfaces should consider incorporating multiple binding motifs that support binding to various integrin and nonintegrin receptors, gradients in ligand density, nanoscale clustering, dynamic interfacial properties, and structural as well as mechanical characteristics of the ECM. For example, recent research indicates that materials with elastic moduli comparable to native tissues and surfaces that direct ECM deposition and assembly up-regulate cellular activities, including proliferation and differentiation [100, 101]. Successful development of these bioactive interfaces will rely heavily on the integration of advances in biochemistry, cell biology, synthetic chemistry, and materials science and engineering.

Acknowledgements AJG gratefully acknowledges support from the National Science Foundation, National Institutes of Health, Arthritis Foundation, Whitaker Foundation, and the Georgia Tech/Emory NSF ERC on Engineering Living Tissues.

References

1. De Arcangelis A, Georges-Labouesse E (2000) Trends Genet 16:389
2. Danen EH, Sonnenberg A (2003) J Pathol 201:632
3. Tanaka E, Sabry J (1995) Cell 83:171
4. Lotz MM, Burdsal CA, Erickson HP, McClay DR (1989) J Cell Biol 109:1795
5. Balaban NQ, Schwarz US, Riveline D, Goichberg P, Tzur G, Sabanay I, Mahalu D, Safran S, Bershadsky A, Addadi L, Geiger B (2001) Nat Cell Biol 3:466
6. Giancotti FG, Ruoslahti E (1999) Science 285:1028
7. Schwartz MA, Assoian RK (2001) J Cell Sci 114:2553
8. Hynes RO (2002) Cell 110:673
9. Wehrle-Haller B, Imhof BA (2003) J Pathol 200:481
10. Jin H, Varner J (2004) Br J Cancer 90:561
11. Anderson JM (2001) Annu Rev Mater Res 31:81
12. Hench LL, Polak JM (2002) Science 295:1014
13. Vreeland WN, Barron AE (2002) Curr Opin Biotechnol 13:87
14. Lutolf MP, Hubbell JA (2005) Nat Biotechnol 23:47
15. Faull RJ, Kovach NL, Harlan J, Ginsberg MH (1993) J Cell Biol 121:155
16. Choquet D, Felsenfeld DP, Sheetz MP (1997) Cell 88:39
17. Geiger B, Bershadsky A, Pankov R, Yamada KM (2001) Nat Rev Mol Cell Biol 2:793
18. Geiger B, Bershadsky A (2002) Cell 110:139
19. Moursi AM, Damsky CH, Lull J, Zimmerman D, Doty SB, Aota S, Globus RK (1996) J Cell Sci 109:1369
20. Moursi AM, Globus RK, Damsky CH (1997) J Cell Sci 110:2187
21. Xiao G, Wang D, Benson MD, Karsenty G, Franceschi RT (1998) J Biol Chem 273:32988
22. Broberg M, Eriksson C, Nygren H (2002) J Lab Clin Med 139:163
23. Gorbet MB, Sefton MV (2003) J Biomed Mater Res A 67:792
24. Tang L, Ugarova TP, Plow EF, Eaton JW (1996) J Clin Invest 97:1329
25. Flick MJ, Du X, Witte DP, Jirouskova M, Soloviev DA, Busuttil SJ, Plow EF, Degen JL (2004) J Clin Invest 113:1596
26. Hu WJ, Eaton JW, Ugarova TP, Tang L (2001) Blood 98:1231
27. García AJ (2005) Biomaterials 26:7525
28. Tang L, Eaton JW (1993) J Exp Med 178:2147
29. Howlett CR, Evans MDM, Walsh WR, Johnson G, Steele JG (1994) Biomaterials 15:213
30. McNally AK, Anderson JM (2002) Am J Pathol 160:621
31. Boyan BD, Hummert TW, Dean DD, Schwartz Z (1996) Biomaterials 17:137
32. Andrade JD, Hlady V (1986) Adv Polym Sci 79:1
33. Hoffman AS (1999) J Biomater Sci Polym Ed 10:1011
34. Agashe M, Raut V, Stuart SJ, Latour RA (2005) Langmuir 21:1103
35. Wilson K, Stuart SJ, Garcia A, Latour RA Jr (2004) J Biomed Mater Res A 69:686
36. Zheng J, Li L, Tsao HK, Sheng YJ, Chen S, Jiang S (2005) Biophys J 89:158
37. Merrill EW (1992) Poly(ethylene oxide) and blood contact: a chronicle of one laboratory. In: Harris JM (ed) Glycol chemistry: biotechnical and biomedical applications. Plenum, New York, p 199
38. Kim JH, Kim SC (2002) Biomaterials 23:2015
39. Norde W, Gage D (2004) Langmuir 20:4162
40. Prime KL, Whitesides GM (1991) Science 252:1164
41. Prime KL, Whitesides GM (1993) J Am Chem Soc 115:10714

42. Harder P, Grunze M, Dahint R, Whitesides GM, Laibinis PE (1998) J Phys Chem B 102:426
43. Luk Y-Y, Kato M, Mrksich M (2000) Langmuir 16:9604
44. Holland NB, Qiu Y, Ruegsegger M, Marchant RE (1998) Nature 392:799
45. Kane RS, Deschatelets P, Whitesides GM (2003) Langmuir 19:2388
46. Keselowsky BG, Collard DM, García AJ (2003) J Biomed Mater Res 66A:247
47. Keselowsky BG, Collard DM, García AJ (2004) Biomaterials 25:5947
48. Keselowsky BG, Collard DM, García AJ (2005) Proc Natl Acad Sci USA 102:5953
49. Lan MA, Gersbach CA, Michael KE, Keselowsky BG, García AJ (2005) Biomaterials 26:4523
50. Langer R, Tirrell DA (2004) Nature 428:487
51. Hersel U, Dahmen C, Kessler H (2003) Biomaterials 24:4385
52. Shin H, Jo S, Mikos AG (2003) Biomaterials 24:4353
53. Garcia AJ, Reyes CD (2005) J Dent Res 84:407
54. Massia SP, Hubbell JA (1991) J Cell Biol 114:1089
55. Maheshwari G, Brown G, Lauffenburger DA, Wells A, Griffith LG (2000) J Cell Sci 113:1677
56. Shin H, Jo S, Mikos AG (2002) J Biomed Mater Res 61:169
57. Schense JC, Hubbell JA (2000) J Biol Chem 275:6813
58. Silva GA, Czeisler C, Niece KL, Beniash E, Harrington DA, Kessler JA, Stupp SI (2004) Science 303:1352
59. Mann BK, West JL (2002) J Biomed Mater Res 60:86
60. Rezania A, Healy KE (2000) J Biomed Mater Res 52:595
61. Rowley JA, Mooney DJ (2002) J Biomed Mater Res 60:217
62. Schense JC, Bloch J, Aebischer P, Hubbell JA (2000) Nat Biotechnol 18:415
63. Ferris DM, Moodie GD, Dimond PM, Gioranni CW, Ehrlich MG, Valentini RF (1999) Biomaterials 20:2323
64. Eid K, Chen E, Griffith L, Glowacki J (2001) J Biomed Mater Res 57:224
65. Alsberg E, Anderson KW, Albeiruti A, Rowley JA, Mooney DJ (2002) Proc Natl Acad Sci USA 99:12025
66. Yu X, Bellamkonda RV (2003) Tissue Eng 9:421
67. Li F, Carlsson D, Lohmann C, Suuronen E, Vascotto S, Kobuch K, Sheardown H, Munger R, Nakamura M, Griffith M (2003) Proc Natl Acad Sci USA 100:15346
68. Huhtala P, Humphries MJ, McCarthy JB, Tremble PM, Werb Z, Damsky CH (1995) J Cell Biol 129:867
69. Sastry SK, Lakonishok M, Thomas DA, Muschler J, Horwitz AF (1996) J Cell Biol 133:169
70. García AJ, Vega MD, Boettiger D (1999) Mol Biol Cell 10:785
71. Mostafavi-Pour Z, Askari JA, Parkinson SJ, Parker PJ, Ng TT, Humphries MJ (2003) J Cell Biol 161:155
72. Tate MC, García AJ, Keselowsky BG, Schumm MA, Archer DR, LaPlaca MC (2004) Mol Cell Neurosci 27:22
73. Scarborough RM, Naughton MA, Teng W, Rose JW, Phillips DR, Nannizzi L, Arfsten A, Campbell AM, Charo IF (1993) J Biol Chem 268:1066
74. Koivunen E, Wang B, Ruoslahti E (1994) J Cell Biol 124:373
75. Humphries JD, Askari JA, Zhang XP, Takada Y, Humphries MJ, Mould AP (2000) J Biol Chem 275:20337
76. García AJ, Schwarzbauer JE, Boettiger D (2002) Biochemistry 41:9063
77. Aota S, Nomizu M, Yamada KM (1994) J Biol Chem 269:24756
78. Redick SD, Settles DL, Briscoe G, Erickson HP (2000) J Cell Biol 149:521

79. Akiyama SK, Aota S, Yamada KM (1995) Cell Adhes Commun 3:13
80. Kao WJ, Lee D, Schense JC, Hubbell JA (2001) J Biomed Mater Res 55:79
81. Dillow AK, Ochsenhirt SE, McCarthy JB, Fields GB, Tirrell M (2001) Biomaterials 22:1493
82. Grant RP, Spitzfaden C, Altroff H, Campbell ID, Mardon HJ (1997) J Biol Chem 272:6159
83. Cutler SM, García AJ (2003) Biomaterials 24:1759
84. Liu JC, Heilshorn SC, Tirrell DA (2004) Biomacromolecules 5:497
85. White DJ, Puranen S, Johnson MS, Heino J (2004) Int J Biochem Cell Biol 36:1405
86. Knight CG, Morton LF, Onley DJ, Peachey AR, Messent AJ, Smethurst PA, Tuckwell DS, Farndale RW, Barnes MJ (1998) J Biol Chem 273:33287
87. Reyes CD, García AJ (2003) J Biomed Mater Res 65A:511
88. Reyes CD, Garcia AJ (2004) J Biomed Mater Res 69A:591
89. Whitesides GM, Ostuni E, Takayama S, Jiang X, Ingber DE (2001) Annu Rev Biomed Eng 3:335
90. Kane RS, Takayama S, Ostuni E, Ingber DE, Whitesides GM (1999) Biomaterials 20:2363
91. Bhatia SN, Yarmush ML, Toner M (1997) J Biomed Mater Res 34:189
92. Singhvi R, Kumar A, Lopez GP, Stephanopoulos GN, Wang DI, Whitesides GM, Ingber DE (1994) Science 264:696
93. Chen CS, Mrksich M, Huang S, Whitesides G, Ingber DE (1997) Science 276:1425
94. McBeath R, Pirone DM, Nelson CM, Bhadriraju K, Chen CS (2004) Dev Cell 6:483
95. Gallant ND, Capadona JR, Frazier AB, Collard DM, García AJ (2002) Langmuir 18:5579
96. Lehnert D, Wehrle-Haller B, David C, Weiland U, Ballestrem C, Imhof BA, Bastmeyer M (2004) J Cell Sci 117:41
97. Flaim CJ, Chien S, Bhatia SN (2005) Nat Methods 2:119
98. Anderson DG, Putnam D, Lavik EB, Mahmood TA, Langer R (2005) Biomaterials 26:4892
99. Gallant ND, Michael KE, García AJ (2005) Mol Biol Cell 16:4329
100. Engler AJ, Griffin MA, Sen S, Bonnemann CG, Sweeney HL, Discher DE (2004) J Cell Biol 166:877
101. Capadona JR, Petrie TA, Fears KP, Latour RA, Collard DM, García AJ (2005) Adv Mater 17:2604

Adv Polym Sci (2006) 203: 191–221
DOI 10.1007/12_070
© Springer-Verlag Berlin Heidelberg 2006
Published online: 5 January 2006

Polymeric Systems for Bioinspired Delivery of Angiogenic Molecules

Claudia Fischbach · David J. Mooney (✉)

Division of Engineering and Applied Sciences, Harvard University,
Room 325 Pierce Hall, 29 Oxford Street, Cambridge, MA 02138, USA
fischbcl@deas.harvard.edu, mooneyd@deas.harvard.edu

Abstract Growth factors are increasingly utilized to promote regeneration of lost or compromised tissues and organs. However, current strategies applying growth factors by bolus injections typically fail to restore tissue functions. Delivery from polymeric systems may overcome this limitation by supplying growth factors in a well-controlled, localized, and sustained manner to the defect site. Traditional polymeric delivery vehicles have been developed based on physicochemical design variables; however, it has now become clear that the appropriate mimicry of certain biologic signaling events may be necessary to achieve full function from the delivered growth factors. Because of its central importance in the development and regeneration of various tissues (e.g., blood vessels, bone, and nerves) bioinspired VEGF supply may be particularly useful to successfully restore tissue functions. Following a brief overview of VEGF's biology, design attributes for polymeric systems for VEGF delivery will be discussed, and subsequently illustrated in the context of three specific applications: therapeutic angiogenesis, bone regeneration, and nerve regeneration.

Keywords Angiogenesis · Drug delivery · Growth factor · Polymer · VEGF

Abbreviations

Ang-1	angiopoietin-1
Ang-2	angiopoietin-2
bFGF	basic fibroblast growth factor
BMP-2	bone morphogenetic protein-2
3-D	three-dimensional
ECM	extracellular matrix
G	alpha-L-guluronic acid
LCST	lower critical solution temperature
M	beta-D-mannuronic acid
MW	molecular weight
MMP	matrix metalloproteinase
NGF	nerve growth factor
NRP	neuropilin
PAA	poly(acrylic acid)
PDGF	platelet derived growth factor
PEG	poly(ethylene glycol)
PEI	poly(ethylenimine)
PGA	poly(glycolic acid)
PLA	poly(lactic acid)
PLGA	poly(lactic-co-glycolic acid)
PNIPAAm	poly(N-isopropylacrylamide)
PPE-EA	polyaminoethyl propylene phosphate
RGD	arginine-glycine-aspartic acid
GRGD	glycine-arginine-aspartic acid
T_g	glass transition temperature
TGF-beta	transforming growth factor-beta
VEGF	vascular endothelial growth factor
VEGF-R	vascular endothelial growth factor receptor
VPF	vascular permeability factor

1
Introduction

Growth factor signaling is key to the sequences of events responsible for both development and regeneration of tissues. Growth factors constitute a complex family of polypeptide molecules exerting versatile biological functions through specific binding to receptors on the cell surface. They are expressed by a variety of different cell types to control cellular migration, proliferation, differentiation, and survival, and ultimately lead to spatially and temporally guided tissue and organ development. Growth factors usually act as local mediators at very low concentrations (about 10^{-9}–10^{-11} M), and control over their signaling activity is mediated through the existence of a biological delivery system. This dynamic system tightly connects growth factor availability, and, thus, signaling activity, to specific cellular needs and involves both intracellular and extracellular control mechanisms [1, 2].

A major aim of medicine is now to generate or regenerate functional tissues to replace lost or compromised tissues and organs, and the rapidly evolving field of tissue engineering increasingly seeks to exploit growth factor signaling pathways to accomplish these goals. The three common strategies presently pursued in tissue regeneration and engineering include (A) conduction (i.e., implantation of biomaterials that provide structural support for ingrowth of the desired healthy host cells), (B) induction (i.e., delivery of growth factors promoting tissue regeneration), and (C) the transplantation of cells capable of participating in tissue regeneration [3, 4]. Growth factors act as potent tissue-inducing substances and may either be administered alone, or in combination with the conductive and cell transplantation approaches.

The goal in growth factor delivery strategies for tissue regeneration is to mimic physiological signaling and achieve biologic functionality. Growth factors restore tissue functions by locally signaling to specific target cell populations. At the same time, signal propagation to more distant, nontarget cells is minimized and this allows for reduction of undesired side effects. Elucidation of cellular proliferation and differentiation cascades furthermore has revealed that isolated signaling of a single growth factor is oftentimes not sufficient for regeneration of functional, mature tissues; but rather simultaneous or sequential cooperation of multiple growth factors may be required for therapeutic efficacy. Temporal and spatial control over the bioavailability of growth factors is critical to all of the above processes. With the aim of reconstituting tissue functions, biomimetic delivery systems may be required to recapitulate these physiological patterns (i.e., to provide growth factors in a controlled localized and sustained fashion).

Because of the central importance of new blood vessel formation in almost all regenerative processes, the design of strategies for delivery of factors that promote the formation of new blood vessels (angiogenesis) is of particular interest. Blood vessels not only provide nutrients and oxygen to cells and remove waste products, but also supply soluble factors and circulating progenitor cells critical to tissue repair. First discovered and termed "vascular permeability factor" (VPF) by Dvorak and co-workers in 1983, and cloned in 1989 by Ferrara and co-workers, VEGF has been intensively examined for its role in blood vessel formation [5, 6]. The central importance of VEGF in development is highlighted by the finding that a 50% reduction in its expression results in embryonic lethality [7, 8]. Since its cloning, recombinant VEGF is available in large quantities, making it an attractive molecule for therapeutic applications.

For almost two decades VEGF was presumed to act specifically on endothelial cells (i.e., the cells that line the lumen of blood vessels). Recent evidence, however, indicates that it can no longer be considered exclusively as an angiogenic growth factor, but must also be considered to be a key regulator of other tissues (e.g., bone and nervous tissue). VEGF indirectly affects many biologic processes as a result of its enhancement of blood vessel ingrowth, which provides for the mass transport requirements and delivery of circulating factors and stem cells participating in tissue repair [9]. Importantly, VEGF also provokes direct effects on nonendothelial cells expressing VEGF receptors, such as smooth muscle cells, chondrocytes, osteoblasts, and neurons [10–14]. VEGF therapy may therefore prove valuable for a much wider range of regenerative applications than originally thought.

Polymeric systems offer great potential as readily controlled carrier systems that allow for spatially and temporally controlled delivery of growth factors. At present, growth factors are most commonly applied in solution form via bolus injection (Fig. 1). However, this delivery route results in unlocalized supply, short tissue exposure times, entails potential problems at nontarget sites such as promotion of diseases, and is expensive because high concentrations of growth factors are typically applied.

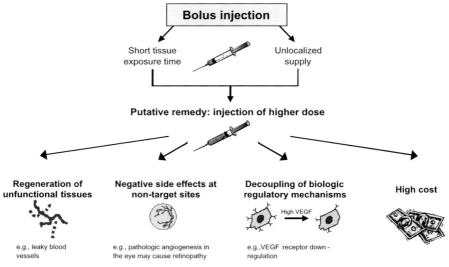

Fig. 1 Currently, growth factors (e.g., VEGF) are most commonly applied via bolus injection. However, this application route typically results in unlocalized supply and short tissue exposure times. In order to overcome these restrictions higher quantities of growth factors may be injected. This approach, however, may still not lead to regeneration of functional tissues but may be associated with negative side effects at nontarget sites, decoupling of the factors providing normal regulation of growth factor signaling (e.g., VEGF receptor downregulation [15]), and higher costs

This article reviews the design and potential importance of polymeric vehicles for bioinspired supply of angiogenic growth factors, and VEGF will be used as a paradigm for the discussion of specific design parameters for bioinspired growth factor delivery. A short overview of VEGF's biology precedes the description of design variables. Currently, therapeutic angiogenesis, bone regeneration, and nerve regeneration represent the best understood areas for VEGF delivery from polymeric systems, and multifactor approaches for these specific applications will be described.

2
VEGF Biology

Angiogenesis is a complex multilevel process regulated by a well-concerted interplay between numerous cell types, proteolytic enzymes, cytokines, and growth factors, and VEGF is one of the most widely studied angiogenic factors [16–18]. It initiates activation, migration, and proliferation of endothelial cells to sprout neovessels (Fig. 2). These newly formed tubes are stabilized through recruitment of and association with mural cells (smooth muscle cells and pericytes) [18–21]. Withdrawal of VEGF prior to stabilization causes regression of nascent vessels due to endothelial cell death. During this sequence of events, VEGF acts in cooperation with other growth factors [18–21]. While fibroblast growth factor (bFGF) and angiopoietin-2 (Ang-2) collaborate in the initiation of the cascade, platelet derived growth factor (PDGF), transforming growth factor-beta (TGF-beta), and angiopoietin-1 (Ang-1) are required in later stages mediating maturation of neovessels by promoting interactions

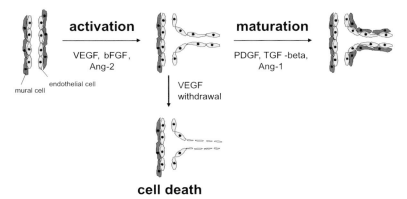

Fig. 2 Angiogenesis is a multistep process involving the synergistic interplay of different growth factors and cell types. VEGF initiates the angiogenic cascade in cooperation with bFGF and Ang-2. Mature and functional blood vessels develop in the presence of PDGF, TGF-beta, and Ang-1; VEGF withdrawal before maturation leads to endothelial cell death

with mural cells (Fig. 2) [18–20]. In order to regenerate functional blood vessels, VEGF may need to be constantly supplied over an extended time period, and in combination with other growth factors.

Polymer systems used for VEGF delivery mimic key aspects by which the extracellular matrix (ECM) of tissues controls the bioavailability and signaling activity of VEGF in the body. Specific characteristics of the ECM regulate the versatile functions of VEGF in tissue regeneration, and provide design criteria for polymeric delivery systems. Specifically, the ECM sequesters VEGF and enables storage of this otherwise rapidly biodegraded factor, presents it in a localized fashion, and enhances the efficiency of signal transduction. Biologic control over VEGF binding to the ECM, and thus its bioavailability, is realized by cellular production of four different isoforms (VEGF121, VEGF165, VEGF189, and VEGF206). A process called alternative splicing generates all four isoforms from the same gene. To what extent the individual isoforms are bound to the ECM is determined by the molecular size of their ECM-binding regions (Fig. 3). VEGF121 completely lacks this part of the molecule and is, therefore, freely diffusible and immediately bioavailable upon secretion from cells. VEGF165, the predominant isoform in the body, exists in both free and bound forms, whereas VEGF189 and VEGF209 are almost completely sequestered, and only liberated on cellular demand (Fig. 3) [22–24]. Cells produce proteolytic, ECM-degrading enzymes (e.g., heparanase or matrix metalloproteinases [MMPs]), and these ultimately trigger the release of soluble, bioactive VEGF from its ECM depots [23, 24]. Furthermore, mechanical stimuli contribute to the release of VEGF from its ECM depots [25].

ECM-binding of VEGF significantly modulates the interactions with its receptors, and consequently, plays an important role in the physiological control of VEGF signal transmission. VEGF receptor-2 (VEGFR-2), currently considered the main receptor, is expressed by a variety of cells, including endothelial cells and nerve cells [14, 22] and transmits signals in cooperation with neuropilin-1 (NRP-1). ECM-binding improves the interactions between

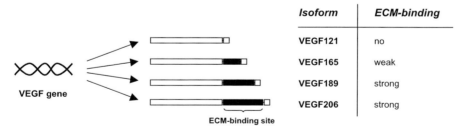

Fig. 3 VEGF exists in four different isoforms that are generated from a single gene and exhibit distinct ECM binding characteristics. Lack of ECM binding allows free diffusion of the VEGF through tissues, and immediate signaling, whereas VEGF sequestered in the ECM represents a depot form that is bioavailable on cellular demand

VEGF, VEGFR-2, and NRP-1, and this ultimately enhances the effectiveness of signal transduction [22, 24, 26]. The importance of ECM components in VEGFR signaling is further emphasized by the finding that VEGF121 appears to elicit reduced biologic potency relative to the ECM-binding isoforms [24].

Biomimicry of the ECM's sustained release properties is critical to the therapeutic success of polymeric systems for VEGF delivery. Specifically, recreation of the physiological ECM binding characteristics may result in ways to stabilize incorporated VEGF, present it in a localized fashion, and improve its signaling capacity.

3
Bioinspired Polymer Delivery Strategies

The main goal of polymeric VEGF delivery systems is controlling the localized and sustained availability of the growth factor. Towards this end, investigation of the physicochemical properties of the delivery vehicle has been the focus of most research to date. Over the last few years, however, an improved understanding of VEGF biology has made it clear that it is not sufficient to only equip VEGF delivery systems with appropriate physicochemical properties, but also that it may be necessary to mimic certain physiological signaling events. Appropriate design of polymeric vehicles in the context of the local biology may result in advanced, bioinspired delivery systems that improve tissue regeneration. In addition, independent from the application, VEGF delivery systems should readily incorporate into the defect site while not compromising the remaining tissue functions or limiting the body's intrinsic regenerative capacity.

3.1
Biological Design Attributes

The properties specifically demanded from the system are prescribed by the characteristics of the tissue to be treated. For example, chronic degenerative diseases such as neurologic diseases and osteoporosis may require continuous delivery of VEGF over a prolonged period of time (months or years) [27], whereas in acute cases of tissue damage (e.g., myocardial infarction or bone fracture) VEGF may prove most beneficial if made bioavailable during a time frame of days or weeks [28, 29]. Furthermore, the spatial dimensions of the devices may vary significantly for different applications. Specifically, slowly progressing, degenerative diseases (e.g., Alzheimer's or Parkinson's disease) not associated with large lesions in early stages of the disease, may be treated with relatively small VEGF delivery devices. In contrast, acute and oftentimes large defects resulting from accidents or tissue resections may require bulkier delivery vehicles that temporarily fill gaps and maintain sufficient space and structural integrity for subsequent tissue formation. Strategies suitable to meet the particular needs of different applications may closely mimic normal VEGF signaling, or be bioinspired approaches that go beyond what normally occurs in the body. Strategies using both concepts may prove particularly beneficial.

3.1.1
Biomimicry of ECM Sequestering Characteristics

Polymer systems that simulate normal ECM-sequestering characteristics may enable control over VEGF temporal and spatial availability, and potentially enhance the effectiveness of VEGF signaling. Delivery of VEGF in association with polymeric ECM mimics may maintain the bioactivity of the growth factor by protecting it from proteolytic degradation and stabilizing the active conformation of the protein. This concept has been supported by the finding that VEGF release from alginate, which exhibits macromolecular properties similar to the natural ECM, leads to greater bioactivity of the factor than direct administration of VEGF [30]. Similarly, VEGF conjugation to fibrin, heparin, or hyaluronan oligosaccharides protects it from clearance, and induces vessel formation more effectively as compared to VEGF alone [31–34]. Incorporation and release of VEGF from polymeric depots in which it is encapsulated have demonstrated maintenance of VEGF bioactivity for up to 30 days [67].

Mechanical stimulation and enzymatic matrix degradation represent typical biologic mechanisms triggering the liberation of VEGF from its ECM depots [24, 25], and their mimicry may allow for localized VEGF supply in concentrations that correspond to the specific cellular demands. Alginate encapsulated VEGF binds in a reversible fashion to the polysaccharide and its release is regulated through mechanical stimuli (Fig. 4) [25]. Implantation of these matrices, followed by mechanical stimulation in vivo, can enhance blood vessel formation, indicating the efficacy and potential clinical utility of mechanically responsive VEGF delivery systems (Fig. 4) [25]. To mimic enzymatically driven VEGF release from ECM stores, polymeric vehicles have been developed that respond to the local activity of proteolytic enzymes (e.g., MMPs, plasmin, heparanase) provided by invading cells [32, 35]. For example, fibrin or peptide cross-linked poly (ethylene glycol) (PEG) gels [31, 35] incorporate VEGF via covalent linkages, or by covalently linking heparin-binding peptides, which then provide affinity sites for VEGF [35–38]. Cellular invasion leads to VEGF release due to the action of ECM degrading proteases [38].

In addition to modifying the VEGF binding characteristics of the polymeric vehicle, one also may exploit the distinct ECM-binding and diffusion characteristics of the different VEGF isoforms. Specifically, VEGF121 is not retained by matrix proteins, and is freely diffusible [24]. Delivery of VEGF121 may allow signaling over great distances. In contrast, delivery of VEGF165,

In vitro   In vivo

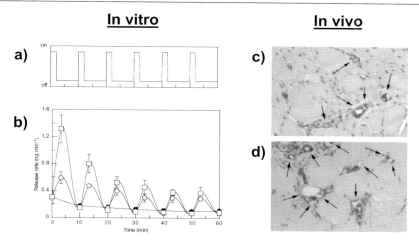

Fig. 4 The release rate of VEGF incorporated into alginate hydrogels is regulated by mechanical strain in vitro and cyclic strain results in enhanced angiogenesis in vivo [25]. Specifically, 6 cycles of mechanical stimulation were performed (mechanical compression for 2 min followed by relaxation for 8 min) and the VEGF release rate monitored in vitro (**a**). Increased amplitude of strain led to increased VEGF release (10% = *open circles*, 25% = *open squares*, no compression = *filled circles*) (**b**). In vivo, mechanical stimulation of alginate gels (**d**) yielded enhanced blood vessel densities relative to un-stimulated gels (**c**). *Arrows* indicate blood vessels in the muscle tissue surrounding the implanted gels (Nature [25], with permission of Nature Publishing Group)

189 and 206 may provide strong, localized signals [24]. Depending on the respective clinical application, combined delivery of VEGF121 along with ECM binding VEGF isoforms may prove beneficial by recreating the favorable functions of both isoforms.

3.1.2
Modulation of Cellular Interactions

The appropriate receipt and transmission of VEGF signaling by cells is affected by their interaction with the surrounding ECM and neighboring cells. Hence, there is considerable interest in controlling the interfacial interactions of polymeric delivery vehicles with the biological system in a way that mimics natural processes. Integrins and cadherins are important families of cell surface receptors that mediate cell–ECM and cell–cell interactions, respectively [39, 40], and coordinately modulate VEGF signaling activity [41–43]. Integrins typically initiate cellular interactions with implanted polymeric devices by recognizing and binding to adsorbed ECM molecules and communicating through the cell membrane into the cells (outside-in signaling). Vice versa, integrins alter cell–ECM interactions in response to intracellular signals (inside-out signaling) [40, 44]. This well-concerted interplay ultimately leads to changes in cell morphology, migration, differentiation, and survival, and is necessary for optimal activation of growth factor receptors (e.g., VEGF receptors) [40–42]. Consequently, integrins can act as master regulators in VEGF signaling.

Delivery systems that promote adhesion of specific cell type(s) (e.g., cells involved in angiogenesis, bone repair, and nerve regeneration) may be required in certain applications of VEGF, specifically those in which it is desirable to mimic integrin signaling events normally occurring during tissue repair. Traditional biomaterials, however, induce nonselective cell adhesion via unspecific protein adsorption, and this may lead to unwanted tissue responses, such as foreign body reactions and fibrous encapsulation of the delivery vehicle [44, 45]. In order to selectively guide cell adhesion, VEGF may be delivered from matrices fabricated from naturally occurring ECM materials (e.g., collagen, fibrin) that inherently provide the desirable integrin ligands [32, 33, 46]. Alternatively, ECM molecules may be used to simply coat the surface of polymer vehicles. However, the binding mediated by secondary forces (e.g., hydrophobic and electrostatic interactions, and hydrogen bonding) is often weak, reversible, and may change the active conformation of the proteins [44, 47]. To increase the quantity of adsorbed ECM molecules physicochemical modifications of the polymer may prove useful [48–50], however, in these approaches the ECM molecules may still be subject to denaturation.

The limitations associated with protein adsorption may effectively be overcome by presenting cell adhesion sites in the form of small immobilized peptides. Many ECM molecules (e.g., fibronectin and collagens) contain the tripeptide arginine-glycine-aspartic acid (RGD) as their cell recognition motif [51, 52], and this peptide sequence may serve as a useful cue to modulate cellular interactions with VEGF delivery systems. Presentation of RGD sequences on otherwise nonadhesive matrices such as alginate and PEG renders these matrices bioactive, and may selectively guide tissue formation in response to VEGF release [35, 53, 54]. The peptides are typically covalently linked to the materials (e.g., between carboxyl groups present on polymers and amino groups on the peptide) by means of bifunctional reagents. Variation of the surface density, the particular type (e.g., GRGD vs. RGD), and the conformation (e.g., cyclic peptides vs. linear) of the utilized sequence facilitates further control over the cell responses (for detailed review see [55]). Alginate gels, for example, have been successfully modified with RGD peptides using carbodiimides chemistry [53, 56–58]. Adjustment of the peptide density and degradation rate of the alginate chains allowed for specific control over cellular adhesion, cell phenotype, and tissue formation, while causing minimal immune response and little capsule formation around the implant [58–60].

In some cases, the polymeric systems may need to effectively prevent cellular adhesion. In particular, VEGF delivery systems in contact with blood (e.g., polymeric coatings of vascular stents) must prevent unspecific adhesion of blood cells, which may otherwise mediate clotting and vessel occlusion [61]. As outlined above, protein adsorption is critical to cellular adhesion and, therefore, PEG has been widely studied for its role in decreasing protein and unspecific cell adhesion [47, 62]. A variety of approaches has been developed to attach PEG to polymer surfaces. For example, synthesis of PEG containing copolymers and covalent grafting of PEG to surfaces that were functionalized with reactive silanol groups proved useful to decrease both protein adsorption and cell attachment [63–65]. Since the nonfouling properties of PEG are dependent on the surface chain density, alternative approaches using plasma deposition of tetraglyme to prepare highly cross-linked PEG-like surfaces have been explored [65].

3.1.3
Multiple Growth Factor Signaling

In the body, VEGF acts in a well-concerted interplay with various other growth factors, and the adequate mimicry of these simultaneous and sequential interactions may be essential to regenerate functional tissues. Simultaneous interactions occur, for example, between VEGF and bFGF during initiation of blood vessel formation [18, 19], and may be recreated by simply incorporating the different proteins into the same polymer delivery system. The utility of this general concept has earlier been demonstrated in the context of bone regeneration and blood vessel formation with growth factor combinations that did not involve VEGF [57, 66]. For example, simultaneous delivery of bone morphogenetic protein-2 (BMP-2) and TGF-beta improved bone formation relative to the individual delivery of either growth factor [57].

Sequential delivery of growth factors has been realized with composite systems that are composed of multiple polymer phases with distinct release kinetics [67, 68]. The growth factor(s) acting early during regeneration are typically incorporated into a rapidly releasing phase, whereas the growth factor(s) signaling later in the process are loaded in a phase with more sustained release characteristics. This approach has been applied in the context of therapeutic angiogenesis for sequential delivery of VEGF and PDGF from poly (lactic-co-glycolic acid) (PLGA) scaffolds [67]. This specific system was designed to release the two growth factors with differential kinetics by mixing polymer microspheres containing pre-encapsulated PDGF with lyophilized VEGF before processing into scaffolds. Since VEGF largely associates with the surface of the scaffold it is subject to rapid release. In contrast, the PDGF incorporation approach results in a more even distribution of factor throughout the matrix, with release regulated by the degradation of the polymer used to form microspheres. In a different approach, sequential delivery of growth factors has been achieved by the development of composite systems consisting of gelatin microspheres incorporated into a synthetic hydrogel matrix (oligo [poly (ethylene glycol) fumarate]) [68].

3.1.4
Gene Transfer

A biologically inspired, but not biomimetic approach for delivery of growth factors, utilizes polymeric delivery of plasmid DNA encoding the desired growth factors (localized gene therapy). Sustained protein delivery is typically controlled by the release properties of the system, but gene transfer strategies additionally rely on the cellular production and secretion of the encoded protein [4]. The key steps in this approach are (A) the release of DNA from its delivery vehicle and cellular uptake of the plasmid and (B) in situ cellular production and secretion of the gene product. The staged process results in a delay in growth factor availability, relative to direct protein delivery. Gene transfer approaches may overcome some of the limitations (e.g., instability within polymers over extended time) associated with direct delivery of proteins [69]. One limitation of this approach is delivery of naked DNA results in low efficiency of transfection, low levels of growth factor production, and factor production for a short time [70]. Condensing DNA with polycationic substances such as poly (ethyleneimine) (PEI) or poly (lysine) to form small positively charged particles prior to delivery can lead to prolonged expression of factors both in vitro and in vivo (Fig. 5) [71–75].

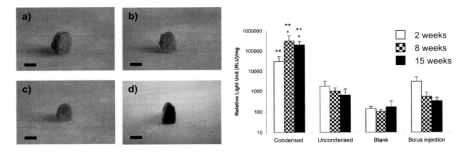

Fig. 5 Release of PEI-condensed plasmid DNA from a porous polymeric scaffold results in prolonged expression of the marker gene beta-galactosidase (**d**) as compared to blank scaffold (**a**), scaffold incorporating plasmid DNA (**b**), and bolus injection of condensed DNA (**c**) [75]. This was shown by macroscopic analysis of X-gal stained scaffolds 15 weeks after subcutaneous implantation (size bars are 3 mm) and by luminescence measurement, which quantitatively assesses gene expression levels (Hum Gene Ther [75], with permission of Mary Ann Liebert, Inc, Publishers)

3.2
Physicochemical Design Variables

Bioinspired VEGF delivery may be implemented by physicochemical strategies that enable temporally and spatially controlled growth factor availability for the desired application, while equipping the respective systems with appropriate physical properties. Relevant physicochemical properties include the rheology and mechanism of gel formation, degradation behavior, and mechanical properties of the system. These features are specified by the polymer and fabrication technique used to form the polymeric vehicle.

3.2.1
Polymers for VEGF Delivery

Synthetic polymers are readily available and exhibit well-defined chemical and physical characteristics allowing for the reproducible fabrication of VEGF delivery systems. Thermoplastic, aliphatic polyesters including poly(glycolic acid) (PGA), poly(lactic acid) (PLA), and their copolymers (PLGA) are the most widely used synthetic polymers for this purpose (Fig. 6). These polyesters are FDA approved and considered biocompatible [76]. When placed in an aqueous environment the otherwise water-insoluble materials degrade through hydrolysis yielding naturally occurring metabolic by-products (lactic and glycolic acid). In order to provide systems that allow for the controlled release of bioactive factors for extended periods of time, these aliphatic polyesters are most commonly processed into microspheres and porous polymer scaffolds [77, 78]. VEGF incorporation into microspheres is typically accomplished with emulsion techniques involving organic solvents, and yields delivery systems that may be readily injected into the body [77, 79, 80]. However, the stability of growth factors incorporated into polymer in this manner may be negatively affected by the organic solvents, and the microclimate in the microspheres may deteriorate the bioactivity of the VEGF [69, 81]. Efforts have been made to overcome these limitations by appropriately modifying the microsphere composition, and developing technologies that do not require organic solvents [69, 82].

Porous polymer scaffolds made from aliphatic polyesters often serve as three-dimensional (3-D) cell carriers in tissue engineering [83–85]. Recently, a gas foaming particulate leaching procedure has been established to allow for scaffold fabrication without the use of the organic solvents or high tempera-

Aliphatic polyesters   **Alginates**

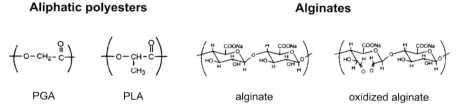

PGA   PLA   alginate   oxidized alginate

Fig. 6 Chemical structure of the aliphatic polyesters poly(glycolic acid) (PGA) and pol(lactic acid) (PLA), sodium alginate, and oxidized sodium alginate

tures commonly applied in other protocols [86]. PLGA scaffolds fabricated using this procedure proved useful as vehicles for sustained VEGF delivery [67, 78], and the VEGF was released in a bioactive form (Fig. 7) [78]. These scaffolds allow for delivery of multiple growth factors in the form of recombinant proteins [67] or plasmid DNA [87].

As an alternative to solid vehicles, VEGF may be delivered from hydrogel-based systems that exhibit structural similarity to the macromolecular-based ECM of many tissues [88]. Importantly, this physical form can typically be introduced into the body using minimally invasive approaches (e.g., injection with syringe). Synthetic materials currently used to form gels for VEGF delivery include derivatives of PEG and poly(acrylic acid) (PAA) [35, 89–92]. PEG hydrogels can be prepared by photopolymerization of macromers modified to contain reactive acrylate termini [89, 90] or via chemical reaction of functionalized PEG macromers with thiol-containing substances [35, 91]. PAA hydrogels based on copolymers containing poly(N-isopropylacrylamide) (PNIPAAm) are attractive VEGF delivery matrices because of their thermoresponsive gelling properties [92]. The lower critical solution temperature (LCST) of PNIPAAm in water, defining the temperature above which gelation occurs, is approximately 32 °C [92]. It can be adjusted to body temperature by copolymerization in order to yield solutions which gel upon injection into the body [88]. Despite offering suitable physical and chemical properties, these synthetic materials may currently be limited in regard to their biocompatibility and biodegradability.

Naturally derived hydrogel forming polymers may provide highly biocompatible vehicles for VEGF delivery. ECM components isolated from tissues and used for VEGF delivery include collagen, hyaluronic acid, and fibrin [36, 93, 94]. Furthermore, chitosan derived from chitin of anthropod exoskeletons has been used for VEGF delivery devices [95, 96]. Alginates rep-

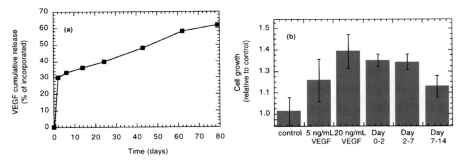

Fig. 7 a The cumulative release of VEGF from PLGA scaffolds fabricated by a gas foaming particulate leaching technique. **b** The bioactivity of VEGF released from these matrices at days 2, 7, and 14 was comparable to the effect obtained with known doses (5 and 20 ng/mL) of VEGF not incorporated into matrices (determined by an endothelial cell proliferation assay) (J. Control. Release [78], with permission of Elsevier)

resent another class of naturally derived materials with potential for VEGF delivery. They are widely used because of their biocompatibility, low toxicity, relatively low cost, and gentle gelling properties [88, 97]. Alginates are linear polysaccharide copolymers extracted from seaweed, and consist of (1-4)-linked beta-D-mannuronic acid (M) and alpha-L-guluronic acid (G) (Fig. 6). Gelation occurs in the presence of divalent cations (e.g., Ca^{2+}), which ionically crosslink the carboxylate groups in the poly-G blocks. Modulation of the crosslinking density by varying the MW of the polymer chains and the M to G ratio yields gels with controlled mechanical properties and pore sizes [97]. One limitation to the use of alginates is their typically slow and unpredictable degradation behavior in the body, which is controlled by the loss of divalent cations, dissolution of the gels, and release of high MW polymer strands that may be difficult for the body to eliminate [88]. To address these restrictions, hydrolytically degradable alginate derivatives have been synthesized by partial oxidation (Fig. 6) and gamma-irradiation [59, 98]. These modified polymers may also be covalently cross-linked with a variety of agents [98, 99]. Gels formed from polymers with a bimodal MW distribution at various oxidation degrees can be formulated to allow decoupling of pre-gel rheological properties from the gel mechanical properties, and control of degradation rate over a wide range [100, 101].

3.2.2
Modulation of Growth Factor Release Kinetics

VEGF release from polymer systems is often controlled by diffusion or polymer degradation, and the desired kinetics are prescribed by the therapeutic application. The highly hydrated, nanoporous gel networks formed by many hydrophilic synthetic or natural polymers that are used for VEGF delivery typically release the factor in a diffusion-controlled manner [30, 88, 97, 102]. The structural integrity and release properties of the swollen networks may be controlled by crosslinking the polymer chains in a covalent or physical fashion. Gels formed by covalently-crosslinked networks are termed "chemical gels", whereas gels generated through secondary forces (e.g., ionic or H-bonding) are denoted as "physical gels" [102]. The rate and distance of VEGF diffusion through these hydrogels is governed by the characteristics of both the network and VEGF. Specifically, the concentration, size, charge, and crosslinking density of the polymer define the nano-porosity of the network and the extent of water absorption [88, 97, 102], whereas the molecular properties of VEGF determine the type and strength of interactions with the polymer. Ionically crosslinked alginate gels represent typical examples of hydrogels that release proteins in a diffusion-controlled manner [30, 103]. If the given hydrodynamic volume of VEGF (roughly correlating with the molecular weight of 45 kDa [24]) is large relative to the pore size of the network VEGF diffusion will be retarded. If one wants to promote cellular infiltration into the delivery vehicle, it is often not desirable to decrease the pore size to regulate VEGF release kinetics, as this may inhibit tissue regeneration [104, 105].

VEGF release from degrading polymeric systems is determined by the rate of degradation and subsequent dissolution [97, 106, 107]. Hydrolysis is frequently exploited as the mechanism to regulate degradation of the polymer system, as this process typically occurs at a consistent rate that is not influenced by local conditions. The molecular weight (MW) distribution and chemical structure of the polymer provide means to readily adjust the VEGF release rate to the specific regenerative needs [108, 109]. For example, delivery systems made from PLGA degrade through hydrolysis of the backbone ester linkages, and the biodegradation rate of PLGA vehicles can be modified on the basis of their MW and molar ratio of lactic and glycolic acid [106, 110–112]. The release rate from vehicles created with naturally derived ECM components that degrade by enzymatic cleavage can also be controlled. For example, growth factor release from collagen, gelatin, and chitosan can depend on enzymatic degradation rather than diffusion [97], and may be decreased by additional crosslinking of the polymer network to slow degradation and dissolution [68, 93, 96, 113].

The desire to control the release of multiple growth factors with distinct kinetics from a single delivery device has motivated the creation of composite systems. A combination of microspheres with a second matrix is an example of these more complex systems. Microspheres may be equipped with differential release properties by fabricating them in different sizes [114], utilizing different natural and synthetic materials (e.g., alginate [56, 95], gelatin [68], collagen [115], and PLGA [67, 77]), and/or retroactively altering them by physicochemical modification (e.g., crosslinking [68, 93] or coating with a second material [95]). The subsequent incorporation of these microspheres into a second matrix (e.g., a synthetic hydrogel matrix [68] or PLGA scaffolds [67]), which may have been equally modified may ultimately lead to a wide spectrum of release kinetics.

3.2.3
Mechanical Properties

Growth factor delivery systems are required to withstand physical forces present at the diseased tissue site. Upon placement into the body, the polymeric vehicles must be able to bear loads and maintain the structural properties critical to the release of the growth factors (e.g., the pore size of hydrogels enabling diffusion) and tissue formation (e.g., provision of space for cellular invasion). At the same time, they often need to effectively transduce biomechanical stimuli to the surrounding cells and tissues. Such signals may be critical to the cellular response to growth factor supply [116–119], and may directly induce cellular proliferation and differentiation [116, 117].

Mechanically competent VEGF delivery systems may be designed on the basis of the estimated mechanical thresholds that the vehicles may encounter in situ.

The mechanical competence of hydrogels is almost entirely controlled by the selected polymer and the type and density of cross-links in the system [120]. Furthermore, the conditions under which the hydrogel network is formed (e.g., temperature, pH, medium), actively influence mechanical performance [97, 120]. The strength of alginate gels, for example, may be raised by increasing the concentration of G residues and the alginate concentration of the polymer [121, 122]. Additionally, the mechanical characteristics of alginate hydrogels may be readily controlled by covalently cross-linking the individual alginate chains with different types and concentrations of crosslinking molecules [123]. Similarly, the mechanical characteristics of other hydrogel matrices such as collagen, chitin, fibrin, and PEG directly correlate with their respective polymer concentration [124–127] and may be further modified by chemical crosslinking [128, 129].

The mechanical strength of solid VEGF delivery systems fabricated from aliphatic polyesters (e.g., PLGA) is prescribed by the polymer MW, crystallinity, and the molar ratio of the individual monomer components (e.g., lactide and glycolide). Matrices made from the homopolymers L-PLA or PGA, as well as copolymers with a high ratio of one monomer are crystalline and exhibit a melting point, whereas those prepared from D,L-PLA and copolymers with less than 90% of one monomer display amorphous properties and undergo glass transition above a critical temperature (T_g) [106, 130]. More crystalline polymers usually exhibit increased physical strength [130]. Since the T_g of the amorphous polymers is typically higher than the body temperature, the respective delivery systems maintain a glassy structure once placed in the body [106, 130, 131], although the absorption of significant amounts of water may decrease T_g to less than 37 °C for certain polymers [132].

The mechanical capacity of polymers is often enhanced by the formation of composite structures composed of (1) oriented reinforcing units, such as fibrils, fibers, or extended chain crystals and (2) the use of binding matrices of the same chemical structure [133]. An alternative that mediates improved mechanical performance while providing space for cellular invasion is the deposition of a bonelike mineral film on the interior pore surfaces of PLGA scaffolds (Fig. 8) [134, 135]. This approach not only leads to increased compressive moduli, but also confers the scaffolds with osteoconductive characteristics and the ability to modulate proliferation and differentiation of multipotent stem cells [134–137].

The size and mechanical characteristics of a delivery system furthermore prescribe its delivery route (i.e., whether it may be surgically inserted into the repair site or injected in a minimally invasive manner). Typically, polymer scaffolds and compact implants are inserted in an invasive fashion, whereas microspheres and gels can be readily applied through injections. The utility of

Fig. 8 Electron micrographs of the surface of **a** a nonmineralized PLG film (molar ratio lactide : glycolide = 85 : 15), or **b** a PLG film with deposited bonelike minerals. **c** Higher magnification image of the mineral film (Biomaterials [136], with permission of Elsevier)

hydrogels as injectable VEGF delivery systems, however, is dependent on the fluid properties of the pre-gelled solution, and the mechanical properties of the post-gel. Specifically, the gel has to be fluid-like during the injection procedure (e.g., flow through a needle, or endoscope), and then transform into a mechanically stable gel upon injection into the body [138]. Alginate hydrogels exhibiting a bimodal MW distribution may represent attractive systems fulfilling these prerequisites, as their fluid and mechanical properties can be decoupled from the total polymer concentration [139]. This finding may be attributed to the fact that low MW alginate components contribute only negligibly to the viscosity of a pre-gelled solution, while they significantly enhance the stiffness of the formed gel; adding a small fraction of high MW alginate to the binary hydrogels additionally maintains a high strain at failure [139]. Consequently, alginate gels with adjusted MW distribution may be readily used as injectable delivery systems for VEGF, and offer favorable properties in mechanically dynamic environments.

4
Applications for VEGF Delivery Systems

VEGF driven tissue regeneration is highly dynamic and involves an elaborate spatiotemporal interplay between cells, the ECM, mechanical signals, and other soluble factors (e.g., growth factors and cytokines). Although these base elements are present in all tissues, the specific characteristics of these components and the kinetics of their interactions may vary according to the precise defect. Traditional polymeric drug carriers may not appropriately mimic the natural complexity of VEGF signaling, whereas bioinspired multifactor approaches have the potential to recapitulate the specific interactions elicited during normal tissue repair (Fig. 9). The diversity of polymeric delivery systems required for various applications will be described below in the context of three specific applications: therapeutic angiogenesis, bone regeneration, and nerve regeneration.

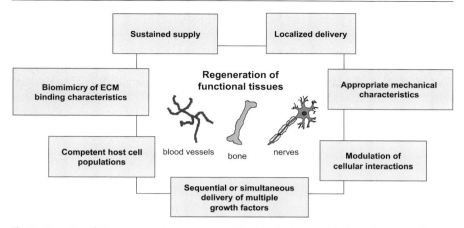

Fig. 9 Functional tissues may be regenerated by designing sophisticated polymeric systems that supply growth factors in a sustained and localized fashion inspired by normal biologic mechanisms, while exhibiting appropriate physicochemical properties

4.1
Therapeutic Angiogenesis

4.1.1
Therapeutic Significance of VEGF

Coronary heart disease, which involves the blockage of coronary arteries and thus reduced blood flow, is the leading cause of death in the United States. It accounts for more than one quarter of all deaths each year. Current treatment options to restore perfusion include lifestyle changes, medication, and invasive surgical procedures [140–142]. However, these approaches are often either restricted to moderately diseased patients or suffer from limitations such as frequent re-narrowing of arteries at later times. The development of alternative strategies may allow better treatment options for patients, and therapeutic angiogenesis, the promotion of blood vessel formation in ischemic tissues, holds great potential as one treatment for cardiovascular occlusive diseases.

Current approaches to therapeutically administer VEGF, either as recombinant protein or by gene transfer, have yielded promising results in animal models and small-scale clinical trials; however, larger trials failed to clearly demonstrate efficacy [16, 143, 144]. In many of these studies recombinant VEGF has been delivered as a single growth factor through bolus injection [143, 145, 146]. Since the elimination half-life of VEGF is approximately 1 h, this approach results in only a temporary supply of the growth factor [147]. Furthermore, spatially confined presentation of growth factors is not achieved with this approach, as highlighted by the finding that injected radio-labeled growth factors yielded target recovery rates of 0.5% after intra-venous injection and 3–5% after injection into the heart [148]. Consequently, inappropriate local and temporal availability of VEGF, and the lack of interaction with other important growth factors likely account for the disappointing results of clinical trials. Enhancing the applied dose may improve the desired outcome at the diseased tissue site; however, this could cause pathologic angiogenesis at nontarget sites (e.g., the eye or dormant tumors) (Fig. 1).

Biomimetic polymeric vehicles may represent an ideal approach to deliver VEGF in a biologically relevant manner. We suggest that appropriately designed systems are capable of regenerating functional blood vessel networks, and that high drug target levels, prolonged presence of the factor, minimized systemic exposure, and simultaneous or sequential supply of multiple growth factors are critical to this end. Specifically, polymeric devices that allow combination therapy of VEGF with vessel stabilizing factors such as PDGF or Ang-1 may lead to enhanced functionality of regenerated blood vessels.

4.1.2
Polymeric Delivery Systems

Polymeric VEGF delivery systems that can be applied in a minimally invasive manner are ideal for therapeutic angiogenesis applications. Acute situations (e.g., infarctions due to occlusion of blood vessels that supply the heart muscle) may require VEGF delivery for weeks, while chronic diseases may require more prolonged VEGF delivery. The physical characteristics of the circulatory system and the heart are amenable to delivery with a vehicle that may be placed by injection through a syringe or catheter, or implantable systems. Hydrogels and microspheres may be suitable candidates for injection as their characteristics can be adjusted according to the local geometry and physical environment (e.g., delivery to either the apex or the base of the heart wall [149]), while porous polymer scaffolds may be useful as patches and wraps.

A variety of different injectable gels and microspheres may be used to enhance local angiogenesis. Delivery from glutaraldehyde cross-linked gelatin microspheres may result in prolonged local presentation of angiogenic factors, relative to bolus injection [150, 151], although the advantage of this carrier over bolus injection has still to be proven [150, 151]. Injectable systems releasing VEGF upon cellular demand have also been prepared from PEG and fibrin matrices [31, 35]. Fibrin inherently promotes blood vessel formation. PEG hydrogel matrices, in contrast, need to be rendered adhesive by polymer conjugation with cell recognition motifs such as RGD-peptides [35, 91]. The concentration of incorporated VEGF can regulate proteolytic enzyme production by endothelial cells, while incorporation of additional growth factors (e.g., TGF-beta) may aid in bioactivation of these enzymes [91], and concomitantly provides signal necessary for the maturation of neovessels [18, 19].

The rheological and mechanical properties of biodegradable, binary alginate systems may be readily adjusted to the physical demands prescribed
by the injection procedure and injection site [139]. Binary gels designed to
exhibit both a high stiffness and a high strain at failure will likely allow prolonged release of VEGF in response to mechanically dynamic environments.
In contrast, less rigid or brittle gels may lead to accelerated release and failure due the strains imposed on the vehicle in the body. Covalent linkage
of cell-adhesion peptides to alginate may allow for control over blood vessel formation by affecting endothelial cell invasion and proliferation [152],
and modulating mural cell properties that stabilize neovessels [153]. Importantly, the gel stiffness may modify the cell phenotype directly and thus VEGF
signaling and blood vessel formation [154]. This characteristic may be particularly useful to improve growth factor production when nonviral VEGF gene
delivery approaches are used to promote vascularization [155]. While hydrogels enable growth factor release over a time-frame of days to approximately
a month [30, 68, 95], microspheres made from aliphatic polyesters may be designed to release drugs over this time frame or longer [110, 130]. However,
long-term release is dependent on growth factor stability in the device [69].

Porous 3-D PLGA scaffolds may be suitable for various therapeutic needs
that benefit from structural support. For example, healing of diabetic ulcers
or enhancing angiogenesis in the lower limbs secondary to diseases such as
diabetes may benefit from the controlled sequential delivery of angiogenic
growth factors from this form of vehicle [67]. These systems additionally provide 3-D support structures that may be readily seeded with cells that can
participate in regeneration. This approach may be particularly useful in the
treatment of patients with a reduced number of competent cells caused, for
example, by aging, disease, or irradiation therapy [156, 157]. In these cases,
the therapeutic efficacy of VEGF delivery may be enhanced by transplantation
of responsive cell populations that are able to actively participate in blood
vessel regeneration (e.g., endothelial cells, bone marrow derived stem cells,
endothelial progenitor cells circulating in the blood) [158–160].

4.2
Bone Regeneration

4.2.1
Therapeutic Significance of VEGF

Significant bone loss caused by accidents, sports injuries, or diseases such as
osteoporosis is frequently associated with severely impaired or nonhealing
defects. Major economic issues arise from an increasingly aged population
that typically exhibits slower healing of fractures, and this leads to a demand
for new therapies to promote bone formation. Bone formation (osteogenesis)
has long been appreciated to be dependent on vascularization, but not until

recently was the critical role of VEGF in this process recognized. After bone injury, VEGF is up-regulated during the first days and gradually decreases in the second week [28]. It directly signals to cells involved in bone formation such as osteoblasts, and mediates their function during bone repair [12, 161–164]. Inhibition of VEGF activity leads to decreased angiogenesis, bone formation, and mineralization [11, 165]. VEGF furthermore interacts with signaling pathways of various osteogenic growth factors [163]. Bone morphogenetic proteins (BMPs) are widely studied for their multifaceted functions in growth and differentiation of bone. Recent studies indicate that their osteogenic potential may be due to synergistic actions with VEGF, and crosstalk between osteoblasts and endothelial cells [166–168].

Polymeric VEGF delivery systems for regeneration of functional bone are required to be mechanically competent, and may also need to appropriately mimic spatial and temporal presentation of multiple growth factors. VEGF supply for two weeks after the initial injury may prove therapeutically beneficial [28]. However, complex and difficult fractures resulting in delayed healing may require a more prolonged application scheme [29]. Synergistic delivery of VEGF with osteogenic growth factors, such as BMPs, may accelerate bone formation and improve the functionality of the regenerated bone. The therapeutic benefit will likely depend on the individual doses, as well as their ratio [169].

4.2.2
Polymeric Delivery Systems

Mineralized PLGA scaffolds provide systems for multiple growth factor delivery that exhibit improved mechanical properties, and offer porous, bone-like substrates that allow for bone conduction [67, 134, 137]. They may be placed during the surgeries frequently required for severe bone fractures. The VEGF release kinetics from these scaffolds is retarded through the mineralization process, and may be further controlled by the polymer composition [170]. Before implantation, these vehicles may be readily seeded with cells that either intrinsically possess bone regenerative properties [136] or that were previously subjected to ex vivo gene therapy to express angiogenic and osteogenic factors (e.g., VEGF and BMPs) [169]. The sustained delivery of condensed plasmid DNA encoding for both VEGF and BMP represents an alternative strategy to yield localized and sustained presentation of these growth factors [87]. Since such approaches rely on the presence of cells that can take up, produce, and secrete the respective factors, simultaneous delivery of competent osteogenic precursor cells may prove beneficial [87]. Furthermore, bone healing and VEGF signaling may be improved by utilizing RGD-modified alginate gels [57, 59, 171], and bone formation with these materials may be regulated by the polymer degradation rate [59]. Providing multiple growth factors [57], and/or incorporating other cell types that can actively partici-

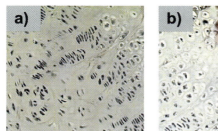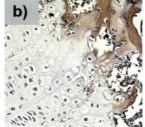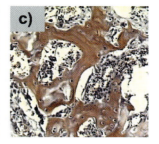

Fig. 10 Cotransplantation of osteoblasts and chondrocytes in RGD-modified alginate enabled formation of growth plate-like structures similar to those in developing long bones. The cellular and tissue morphology at the interface between cartilaginous and bony regions demonstrated cartilage (**a**), transition (**b**), and bone and marrow space (**c**) typical of the corresponding regions in a growth plate (magnification: 200×) (Proc Natl Acad Sci USA [171], with permission of National Academy of Sciences)

pate in bone regeneration (e.g., osteoblasts, bone marrow derived stem cells, chondrocytes) may also be beneficial (Fig. 10) [57, 59, 171].

4.3
Nerve Regeneration

4.3.1
Therapeutic Significance of VEGF

Severe nerve damage, as a consequence of injuries or degenerative disorders such as Alzheimer's or Parkinson's disease, affects millions of people in the prime of their life, and is often incurable, disabling, and fatal. VEGF has recently been recognized to play a fundamental role in nerve regeneration by both providing growth and survival signals to neural cells, and enhancing vascularization. Following nerve damage, VEGF is up-regulated and activates proliferation, migration, and survival of various nerve cells, including neurons and glial cells [27, 172–174]. Concomitantly, VEGF promotes blood vessel ingrowth and ultimately guides the branching, differentiation, and overall function of nerves. A synergistic interplay of VEGF and other growth factors, as well as crosstalk between nerve cells and endothelial cells, are critical to this end. Nerve growth factor (NGF) was the first identified and best characterized neurotrophic factor. Recent evidence suggests that NGF elicits an angiogenic response from cells, and that its neurotrophic capacity may at least partly be linked to this ability [175–177]. Despite the therapeutic potential of NGF, early attempts to develop clinical treatments with this factor were not successful [178]. The inability of these approaches to mimic biologic events, such as combined presentation of VEGF and NGF, may be partially responsible for the failure of the trials.

Development of biomimetic delivery strategies that provide VEGF locally, and in an appropriately timed fashion offer great potential for recreation of damaged nerve functions. Recent data indicate that chronic delivery of VEGF may delay progression of brain-related degenerative diseases [27]. Systemic VEGF delivery would prove ineffective for these applications because of its inability to cross the blood-brain barrier, and implantation of sustained release vehicles into the brain may overcome this issue and other limitations of bolus delivery. Combined delivery of VEGF with neurotrophic factors such as NGF may further improve the functions of regenerated nerves.

4.3.2
Polymeric Delivery Systems

Polymeric guidance channels, microspheres, and hydrogels may all be useful for long-term release of VEGF. A widely explored strategy to treat peripheral nerve and spinal cord injuries is to physically guide neuronal outgrowth by utilizing conduits typically prepared from biodegradable polymers (e.g., PLGA [179, 180] or magnetically aligned collagen or fibrin gels [181, 182]). These grafts are placed across lesions and may exhibit either single or multiple lumens, and serve to support axonal outgrowth, allow the generation of growth factor diffusion gradients, and minimize the infiltration of scar tissue [183]. The release of growth factors from these grafts may render them more effective, as shown with PLGA conduits that released NGF [184]. By varying the approach to factor incorporation, and the PLGA composition, different release kinetics could be readily achieved.

Microspheres are versatile, injectable systems for delivery of neurotrophic growth factors, and the release kinetics with this system may be readily tailored. Long-term release of neurotrophic growth factors has, for example, been achieved with PLGA microspheres [185–187], and alginate and chitosan microspheres may also prove valuable for this application [188]. Furthermore, microspheres may be used in composite systems for multiple growth factor delivery, and incorporated into hydrogel systems that physically direct growth of regenerating nerves. The latter approach yields contact guiding neurotrophic matrices [186], whose effectiveness may potentially be improved by incorporation of additional microspheres that release synergistically acting growth factors such as VEGF.

The performance of polymeric VEGF delivery systems may be further improved by modulation of cellular interactions, transplantation of cells, and delivery of condensed DNA encoding neurotrophic growth factors. Axonal outgrowth may be controlled by providing interfaces that present nerve cell specific recognition motifs, and the modification of hydrogels (e.g., alginate, PEG, fibrin) with integrin ligands known to be involved in nerve regeneration may prove useful [189–191]. These hydrogels may either be used alone, or as a filling of solid nerve conduits. Cellular therapies may particularly aid patients with decreased neural cell signaling activity due to deficits in VEGF supply, and may utilize both mature neural cell types (e.g., glial cells, olfactory ensheathing cells) and progenitor cells (e.g., neural stem cells, glial progenitor cells) [180, 183, 187]. For example, neural stem cells can be seeded onto appropriately designed PLGA scaffolds that emulate their spatial arrangement into the gray and white matter of the spinal cord, and implantation of these constructs mediated functional recovery subsequent to spinal cord injuries [180]. In the future, nonviral gene therapy approaches may become an option to obtain prolonged presentation of neurotrophic factors, and could be achieved with localized delivery of polycation/plasmid DNA complexes to damaged nerve tissue. For example, polycationic polyaminoethyl propylene phosphate (PPE-EA) may mediate gene transfer in the brain, and display prolonged gene expression, lower cytotoxicity, and better nervous tissue compatibility as compared to other condensing agents [192].

5
Summary and Future Directions

Current strategies of growth factor delivery for tissue regeneration most commonly rely on bolus injections, but this approach typically fails to restore compromised tissue functions. Delivery from polymeric systems may overcome the limitations associated with bolus delivery by supplying the growth factor in a well-controlled, localized, and sustained manner to the defect site. Polymeric delivery vehicles have traditionally been developed by modulating physicochemical characteristics. However, it has now become clear that the appropriate mimicry of certain biologic signaling events is useful to achieve full function from the delivered growth factors.

Originally examined solely for its role in blood vessel formation, VEGF has recently been attributed potential in other tissue regeneration therapies. At present, therapeutic angiogenesis, bone regeneration, and nerve regeneration represent the best understood application areas for VEGF delivery from polymeric delivery systems. In the body, VEGF is delivered from a sophisticated, biologic delivery system, whose functionalities may be mimicked by the polymer systems to obtain increased signaling activities from the delivered VEGF. Simulation of the ECM binding and sequestering of VEGF by polymer systems may not only enable control over the temporal and spatial availability of VEGF, but also enhance the effectiveness of signal transduction. ECM molecules furthermore contain cell recognition motifs that are critical to both the cellular interactions with the delivery system and the signaling functions of VEGF. Incorporation of these motifs into the polymer delivery vehicle may more appropriately prime host cells to the delivered factor. A key issue will likely be to present VEGF in the context of other growth factors, as the concerted sequential interplay of multiple factors is critical to blood vessel formation, and tissue regeneration more broadly.

Finally, normal tissue repair is based on elaborate spatiotemporal interactions between the growth factors, cells, the host ECM, and mechanical signals, and it may be necessary to create polymer systems that enable the interplay of these multi-components in an appropriate manner. Depending on the affected tissue type (e.g., blood vessels vs. bone), the patient (e.g., aged patients with fewer competent cells vs. young patients), and the form of tissue damage (e.g., acute injury or degenerative disease), different strategies of VEGF delivery and different polymeric systems may be needed.

References

1. Chen RR, Mooney DJ (2003) Pharm Res 20:1103
2. Alberts B, Johnson A, Lewis J, Raff M, Roberts K, Walter P (2002) Cell communication. In: Gibbs S (ed) Molecular Biology of the Cell. Garland Science, New York, chap 15, p 831
3. Alsberg E, Hill EE, Mooney DJ (2001) Crit Rev Oral Biol Med 12:64
4. Leach JK, Mooney DJ (2004) Expert Opin Biol Ther 4:1015
5. Senger DR, Galli SJ, Dvorak AM, Perruzzi CA, Harvey VS, Dvorak HF (1983) Science 219:983
6. Ferrara N, Henzel WJ (1989) Biochem Biophys Res Commun 161:851
7. Ferrara N, Carver-Moore K, Chen H, Dowd M, Lu L, O'Shea KS, Powell-Braxton L, Hillan KJ, Moore MW (1996) Nature 380:439
8. Carmeliet P, Ferreira V, Breier G, Pollefeyt S, Kieckens L, Gertsenstein M, Fahrig M, Vandenhoeck A, Harpal K, Eberhardt C, Declercq C, Pawling J, Moons L, Collen D, Risau W, Nagy A (1996) Nature 380:435
9. Roufosse CA, Direkze NC, Otto WR, Wright NA (2004) Int J Biochem Cell Biol 36:585
10. Ishida A, Murray J, Saito Y, Kanthou C, Benzakour O, Shibuya M, Wijelath ES (2001) J Cell Physiol 188:359
11. Gerber HP, Vu TH, Ryan AM, Kowalski J, Werb Z, Ferrara N (1999) Nat Med 5:623
12. Mayr-Wohlfart U, Waltenberger J, Hausser H, Kessler S, Gunther KP, Dehio C, Puhl W, Brenner RE (2002) Bone 30:472
13. Carlevaro MF, Cermelli S, Cancedda R, Descalzi CF (2000) J Cell Sci 113(Pt1):59
14. Khaibullina AA, Rosenstein JM, Krum JM (2004) Brain Res Dev Brain Res 148:59
15. Duval M, Bedard-Goulet S, Delisle C, Gratton JP (2003) J Biol Chem 278:20091
16. Ferrara N, Alitalo K (1999) Nat Med 5:1359
17. Ferrara N (1999) Kidney Int 56:794
18. Hirschi KK, Skalak TC, Peirce SM, Little CD (2002) Ann NY Acad Sci 961:223
19. Ennett AB, Mooney DJ (2002) Expert Opin Biol Ther 2:805
20. Carmeliet P (2000) Nat Med 6:389
21. Yancopoulos GD, Davis S, Gale NW, Rudge JS, Wiegand SJ, Holash J (2000) Nature 407:242
22. Cross MJ, Dixelius J, Matsumoto T, Claesson-Welsh L (2003) Trends Biochem Sci 28:488
23. Ferrara N, Davis-Smyth T (1997) Endocr Rev 18:4

24. Ferrara N, Gerber HP, LeCouter J (2003) Nat Med 9:669
25. Lee KY, Peters MC, Anderson KW, Mooney DJ (2000) Nature 408:998
26. Soker S, Takashima S, Miao HQ, Neufeld G, Klagsbrun M (1998) Cell 92:735
27. Storkebaum E, Lambrechts D, Dewerchin M, Moreno-Murciano MP, Appelmans S, Oh H, Van Damme P, Rutten B, Man WY, De Mol M, Wyns S, Manka D, Vermeulen K, Van Den BL, Mertens N, Schmitz C, Robberecht W, Conway EM, Collen D, Moons L, Carmeliet P (2005) Nat Neurosci 8:85
28. Uchida S, Sakai A, Kudo H, Otomo H, Watanuki M, Tanaka M, Nagashima M, Nakamura T (2003) Bone 32:491
29. Kokubu T, Hak DJ, Hazelwood SJ, Reddi AH (2003) J Orthop Res 21:503
30. Peters MC, Isenberg BC, Rowley JA, Mooney DJ (1998) J Biomater Sci Polym Ed 9:1267
31. Ehrbar M, Djonov VG, Schnell C, Tschanz SA, Martiny-Baron G, Schenk U, Wood J, Burri PH, Hubbell JA, Zisch AH (2004) Circ Res 94:1124
32. Ehrbar M, Metters A, Zammaretti P, Hubbell JA, Zisch AH (2005) J Control Release 101:93
33. Steffens GC, Yao C, Prevel P, Markowicz M, Schenck P, Noah EM, Pallua N (2004) Tissue Eng 10:1502
34. Montesano R, Kumar S, Orci L, Pepper MS (1996) Lab Invest 75:249
35. Zisch AH, Lutolf MP, Ehrbar M, Raeber GP, Rizzi SC, Davies N, Schmokel H, Bezuidenhout D, Djonov V, Zilla P, Hubbell JA (2003) FASEB J 17:2260
36. Zisch AH, Schenk U, Schense JC, Sakiyama-Elbert SE, Hubbell JA (2001) J Control Release 72:101
37. Sakiyama-Elbert SE, Hubbell JA (2000) J Control Release 65:389
38. Zisch AH, Lutolf MP, Hubbell JA (2003) Cardiovasc Pathol 12:295
39. Patel SD, Chen CP, Bahna F, Honig B, Shapiro L (2003) Curr Opin Struct Biol 13:690
40. Giancotti FG, Ruoslahti E (1999) Science 285:1028
41. Zachary I, Gliki G (2001) Cardiovasc Res 49:568
42. Kaneko Y, Kitazato K, Basaki Y (2004) J Cell Sci 117:407
43. Zanetti A, Lampugnani MG, Balconi G, Breviario F, Corada M, Lanfrancone L, Dejana E (2002) Arterioscler Thromb Vasc Biol 22:617
44. Ratner BD, Bryant SJ (2004) Annu Rev Biomed Eng 6:41
45. Anderson JM (2001) Annu Rev Mater Res 31:81
46. Kanematsu A, Yamamoto S, Ozeki M, Noguchi T, Kanatani I, Ogawa O, Tabata Y (2004) Biomaterials 25:4513
47. Wilson CJ, Clegg RE, Leavesley DI, Pearcy MJ (2005) Tissue Eng 11:1
48. Zhu H, Ji J, Tan Q, Barbosa MA, Shen J (2003) Biomacromolecules 4:378
49. Ma Z, Gao C, Gong Y, Shen J (2005) Biomaterials 26:1253
50. Tan WJ, Teo GP, Liao K, Leong KW, Mao HQ, Chan V (2005) Biomaterials 26:891
51. Ruoslahti E, Pierschbacher MD (1987) Science 238:491
52. Hersel U, Dahmen C, Kessler H (2003) Biomaterials 24:4385
53. Rowley JA, Madlambayan G, Mooney DJ (1999) Biomaterials 20:45
54. Burdick JA, Anseth KS (2002) Biomaterials 23:4315
55. Hersel U, Dahmen C, Kessler H (2003) Biomaterials 24:4385
56. Loebsack A, Greene K, Wyatt S, Culberson C, Austin C, Beiler R, Roland W, Eiselt P, Rowley J, Burg et al. (2001) J Biomed Mater Res 57:575
57. Simmons CA, Alsberg E, Hsiong S, Kim WJ, Mooney DJ (2004) Bone 35:562
58. Rowley JA, Mooney DJ (2002) J Biomed Mater Res 60:217
59. Alsberg E, Kong HJ, Hirano Y, Smith MK, Albeiruti A, Mooney DJ (2003) J Dent Res 82:903

60. Halberstadt C, Austin C, Rowley J, Culberson C, Loebsack A, Wyatt S, Coleman S, Blacksten L, Burg K, Mooney et al. (2002) Tissue Eng 8:309
61. Herrmann R, Schmidmaier G, Markl B, Resch A, Hahnel I, Stemberger A, Alt E (1999) Thromb Haemost 82:51
62. Hayman EG, Pierschbacher MD, Suzuki S, Ruoslahti E (1985) Exp Cell Res 160:245
63. Lucke A, Tessmar J, Schnell E, Schmeer G, Gopferich A (2000) Biomaterials 21:2361
64. Suggs LJ, West JL, Mikos AG (1999) Biomaterials 20:683
65. Alcantar NA, Aydil ES, Israelachvili JN (2000) J Biomed Mater Res 51:343
66. Marui A, Kanematsu A, Yamahara K, Doi K, Kushibiki T, Yamamoto M, Itoh H, Ikeda T, Tabata Y, Komeda M (2005) J Vasc Surg 41:82
67. Richardson TP, Peters MC, Ennett AB, Mooney DJ (2001) Nat Biotechnol 19:1029
68. Holland TA, Tabata Y, Mikos AG (2005) J Control Release 101:111
69. Zhu G, Mallery SR, Schwendeman SP (2000) Nat Biotechnol 18:52
70. Laham RJ, Simons M, Sellke F (2001) Annu Rev Med 52:485
71. Boussif O, Lezoualc'h F, Zanta MA, Mergny MD, Scherman D, Demeneix B, Behr JP (1995) Proc Natl Acad Sci USA 92:7297
72. Godbey WT, Barry MA, Saggau P, Wu KK, Mikos AG (2000) J Biomed Mater Res 51:321
73. Huang YC, Connell M, Park Y, Mooney DJ, Rice KG (2003) J Biomed Mater Res A 67:1384
74. Huang YC, Simmons C, Kaigler D, Rice KG, Mooney DJ (2005) Gene Ther 12:418
75. Huang YC, Riddle K, Rice KG, Mooney DJ (2005) Hum Gene Ther 16:609
76. Jain R, Shah NH, Malick AW, Rhodes CT (1998) Drug Dev Ind Pharm 24:703
77. Cleland JL, Duenas ET, Park A, Daugherty A, Kahn J, Kowalski J, Cuthbertson A (2001) J Control Release 72:13
78. Sheridan MH, Shea LD, Peters MC, Mooney DJ (2000) J Control Release 64:91
79. Faranesh AZ, Nastley MT, Perez dlC, Haller MF, Laquerriere P, Leong KW, McVeigh ER (2004) Magn Reson Med 51:1265
80. King TW, Patrick CW Jr (2000) J Biomed Mater Res 51:383
81. Lucke A, Gopferich A (2003) Eur J Pharm Biopharm 55:27
82. Schwach G, Oudry N, Delhomme S, Luck M, Lindner H, Gurny R (2003) Eur J Pharm Biopharm 56:327
83. Langer R, Vacanti JP (1993) Science 260:920
84. Widmer MS, Mikos AG (1998) Fabrication of biodegradable polymer scaffolds for tissue engineering. In: Patrick CW Jr, Mikos AG, McIntire LV (eds) Frontiers in tissue engineering. Elsevier Science Ltd, Oxford, p 107
85. Mooney DJ, Mikos AG (1999) Sci Am 280:60
86. Harris LD, Kim BS, Mooney DJ (1998) J Biomed Mater Res 42:396
87. Huang YC, Kaigler D, Rice KG, Krebsbach PH, Mooney DJ (2005) J Bone Miner Res 20:848
88. Lee KY, Mooney DJ (2001) Chem Rev 101:1869
89. Cruise GM, Scharp DS, Hubbell JA (1998) Biomaterials 19:1287
90. West JL, Hubbell JA (1995) Biomaterials 16:1153
91. Seliktar D, Zisch AH, Lutolf MP, Wrana JL, Hubbell JA (2004) J Biomed Mater Res A 68:704
92. Kavanagh CA, Gorelova TA, Selezneva II, Rochev YA, Dawson KA, Gallagher WM, Gorelov AV, Keenan AK (2005) J Biomed Mater Res A 72A:25
93. Tabata Y, Miyao M, Ozeki M, Ikada Y (2000) J Biomater Sci Polym Ed 11:915
94. Peattie RA, Nayate AP, Firpo MA, Shelby J, Fisher RJ, Prestwich GD (2004) Biomaterials 25:2789

95. Lee KW, Yoon JJ, Lee JH, Kim SY, Jung HJ, Kim SJ, Joh JW, Lee HH, Lee DS, Lee SK (2004) Transplant Proc 36:2464
96. Ishihara M, Obara K, Ishizuka T, Fujita M, Sato M, Masuoka K, Saito Y, Yura H, Matsui T, Hattori H, Kikuchi M, Kurita A (2003) J Biomed Mater Res A 64:551
97. Drury JL, Mooney DJ (2003) Biomaterials 24:4337
98. Bouhadir KH, Lee KY, Alsberg E, Damm KL, Anderson KW, Mooney DJ (2001) Biotechnol Prog 17:945
99. Bouhadir KH, Hausman DS, Mooney DJ (1999) Polymer 40:3575
100. Boontheekul T, Kong HJ, Mooney DJ (2005) Biomaterials 26:2455
101. Kong HJ, Kaigler D, Kim K, Mooney DJ (2004) Biomacromolecules 5:1720
102. Hoffman AS (2002) Adv Drug Delivery Rev 54:3
103. Elcin YM, Dixit V, Gitnick G (2001) Artif Organs 25:558
104. Eiselt P, Yeh J, Latvala RK, Shea LD, Mooney DJ (2000) Biomaterials 21:1921
105. Itala AI, Ylanen HO, Ekholm C, Karlsson KH, Aro HT (2001) J Biomed Mater Res 58:679
106. Jain RA (2000) Biomaterials 21:2475
107. Gopferich A (1996) Biomaterials 17:103
108. Saito N, Okada T, Horiuchi H, Murakami N, Takahashi J, Nawata M, Ota H, Nozaki K, Takaoka K (2001) Nat Biotechnol 19:332
109. Alonso MJ, Gupta RK, Min C, Siber GR, Langer R (1994) Vaccine 12:299
110. Freiberg S, Zhu XX (2004) Int J Pharm 282:1
111. Park TG (1994) J Control Release 30:161
112. Makino K, Mogi T, Ohtake N, Yoshida M, Ando S, Nakajima T, Ohshima H (2000) Colloids Surf B Biointerfaces 19:173
113. Ozeki M, Ishii T, Hirano Y, Tabata Y (2001) J Drug Target 9:461
114. Berkland C, King M, Cox A, Kim K, Pack DW (2002) J Control Release 82:137
115. Kanematsu A, Marui A, Yamamoto S, Ozeki M, Hirano Y, Yamamoto M, Ogawa O, Komeda M, Tabata Y (2004) J Control Release 99:281
116. Kim BS, Nikolovski J, Bonadio J, Mooney DJ (1999) Nat Biotechnol 17:979
117. Nikolovski J, Kim BS, Mooney DJ (2003) FASEB J 17:455
118. Butler DL, Goldstein SA, Guilak F (2000) J Biomech Eng 122:570
119. Sikavitsas VI, Temenoff JS, Mikos AG (2001) Biomaterials 22:2581
120. Anseth KS, Bowman CN, Brannon-Peppas L (1996) Biomaterials 17:1647
121. Draget KI, Skjak-Braek G, Smidsrod O (1997) Int J Biol Macromol 21:47
122. LeRoux MA, Guilak F, Setton LA (1999) J Biomed Mater Res 47:46
123. Lee KY, Rowley JA, Eiselt P, Moy E, Bouhadir KH, Mooney DJ (2000) Macromolecules 33:4291
124. Krishnan L, Weiss JA, Wessman MD, Hoying JB (2004) Tissue Eng 10:241
125. Gerentes P, Vachoud L, Doury J, Domard A (2002) Biomaterials 23:1295
126. Benkherourou M, Gumery PY, Tranqui L, Tracqui P (2000) IEEE Trans Biomed Eng 47:1465
127. Bryant SJ, Anseth KS (2002) J Biomed Mater Res 59:63
128. Charulatha V, Rajaram A (2003) Biomaterials 24:759
129. Bryant SJ, Anseth KS, Lee DA, Bader DL (2004) J Orthop Res 22:1143
130. Saltzman WM (2001) Overview of polymeric materials. In: Saltzman WM (ed) Drug Delivery – Engineering principles for drug therapy. Oxford University Press, New York, chap Appendix A, p 320
131. Jamshidi K, Hyon S-H, Ikada Y (1988) Polymer 29:2229
132. Schliecker G, Schmidt C, Fuchs S, Wombacher R, Kissel T (2003) Int J Pharm 266:39

133. Ashammakhi N, Peltoniemi H, Waris E, Suuronen R, Serlo W, Kellomaki M, Tormala P, Waris T (2001) Plast Reconstr Surg 108:167
134. Murphy WL, Kohn DH, Mooney DJ (2000) J Biomed Mater Res 50:50
135. Murphy WL, Mooney DJ (2002) J Am Chem Soc 124:1910
136. Murphy WL, Hsiong S, Richardson TP, Simmons CA, Mooney DJ (2005) Biomaterials 26:303
137. Murphy WL, Simmons CA, Kaigler D, Mooney DJ (2004) J Dent Res 83:204
138. Kim BS, Mooney DJ (1998) Trends Biotechnol 16:224
139. Kong HJ, Lee KY, Mooney DJ (2002) Polymer 43:6239
140. Hansson GK (2005) N Engl J Med 352:1685
141. Taggart DP (2005) BMJ 330:785
142. Sleight P (2003) Am J Cardiol 92:4
143. Epstein SE, Fuchs S, Zhou YF, Baffour R, Kornowski R (2001) Cardiovasc Res 49:532
144. Yoon YS, Johnson IA, Park JS, Diaz L, Losordo DW (2004) Mol Cell Biochem 264:63
145. Post MJ, Laham R, Sellke FW, Simons M (2001) Cardiovasc Res 49:522
146. Syed IS, Sanborn TA, Rosengart TK (2004) Cardiology 101:131
147. Lazarous DF, Shou M, Scheinowitz M, Hodge E, Thirumurti V, Kitsiou AN, Stiber JA, Lobo AD, Hunsberger S, Guetta E, Epstein SE, Unger EF (1996) Circulation 94:1074
148. Lazarous DF, Shou M, Stiber JA, Dadhania DM, Thirumurti V, Hodge E, Unger EF (1997) Cardiovasc Res 36:78
149. Guccione JM, Costa KD, McCulloch AD (1995) J Biomech 28:1167
150. Iwakura A, Fujita M, Kataoka K, Tambara K, Sakakibara Y, Komeda M, Tabata Y (2003) Heart Vessels 18:93
151. Sakakibara Y, Tambara K, Sakaguchi G, Lu F, Yamamoto M, Nishimura K, Tabata Y, Komeda M (2003) Eur J Cardiothorac Surg 24:105
152. Sagnella SM, Kligman F, Anderson EH, King JE, Murugesan G, Marchant RE, Kottke-Marchant K (2004) Biomaterials 25:1249
153. Mann BK, Gobin AS, Tsai AT, Schmedlen RH, West JL (2001) Biomaterials 22:3045
154. Kong HJ, Polte TR, Alsberg E, Mooney DJ (2005) Proc Natl Acad Sci USA 102:4300
155. Kong HJ, Liu J, Riddle K, Matsumoto T, Leach K, Mooney DJ (2005) Nat Mater 4:460
156. Stenderup K, Justesen J, Clausen C, Kassem M (2003) Bone 33:919
157. Jacobsson MG, Jonsson AK, Albrektsson TO, Turesson IE (1985) Plast Reconstr Surg 76:841
158. Peters MC, Polverini PJ, Mooney DJ (2002) J Biomed Mater Res 60:668
159. Nor JE, Peters MC, Christensen JB, Sutorik MM, Linn S, Khan MK, Addison CL, Mooney DJ, Polverini PJ (2001) Lab Invest 81:453
160. Nor JE, Christensen J, Mooney DJ, Polverini PJ (1999) Am J Pathol 154:375
161. Deckers MM, Karperien M, van Der BC, Yamashita T, Papapoulos SE, Lowik CW (2000) Endocrinology 141:1667
162. Midy V, Plouet J (1994) Biochem Biophys Res Commun 199:380
163. Carano RA, Filvaroff EH (2003) Drug Discov Today 8:980
164. Gerber HP, Ferrara N (2000) Trends Cardiovasc Med 10:223
165. Street J, Bao M, deGuzman L, Bunting S, Peale FV Jr, Ferrara N, Steinmetz H, Hoeffel J, Cleland JL, Daugherty A, van Bruggen N, Redmond HP, Carano RA, Filvaroff EH (2002) Proc Natl Acad Sci USA 99:9656
166. Deckers MM, van Bezooijen RL, van der HG, Hoogendam J, van Der BC, Papapoulos SE, Lowik CW (2002) Endocrinology 143:1545
167. Bouletreau PJ, Warren SM, Spector JA, Peled ZM, Gerrets RP, Greenwald JA, Longaker MT (2002) Plast Reconstr Surg 109:2384

168. Kaigler D, Krebsbach PH, West ER, Horger K, Huang YC, Mooney DJ (2005) FASEB J 19:665
169. Peng H, Wright V, Usas A, Gearhart B, Shen HC, Cummins J, Huard J (2002) J Clin Invest 110:751
170. Murphy WL, Peters MC, Kohn DH, Mooney DJ (2000) Biomaterials 21:2521
171. Alsberg E, Anderson KW, Albeiruti A, Rowley JA, Mooney DJ (2002) Proc Natl Acad Sci USA 99:12025
172. Sondell M, Lundborg G, Kanje M (1999) J Neurosci 19:5731
173. Silverman WF, Krum JM, Mani N, Rosenstein JM (1999) Neuroscience 90:1529
174. Storkebaum E, Lambrechts D, Carmeliet P (2004) Bioessays 26:943
175. Cantarella G, Lempereur L, Presta M, Ribatti D, Lombardo G, Lazarovici P, Zappala G, Pafumi C, Bernardini R (2002) FASEB J 16:1307
176. Calza L, Giardino L, Giuliani A, Aloe L, Levi-Montalcini R (2001) Proc Natl Acad Sci USA 98:4160
177. Moser KV, Reindl M, Blasig I, Humpel C (2004) Brain Research 1017:53
178. Thoenen H, Sendtner M (2002) Nat Neurosci 5 Suppl:1046
179. Evans GR, Brandt K, Niederbichler AD, Chauvin P, Herrman S, Bogle M, Otta L, Wang B, Patrick CW Jr (2000) J Biomater Sci Polym Ed 11:869
180. Teng YD, Lavik EB, Qu X, Park KI, Ourednik J, Zurakowski D, Langer R, Snyder EY (2002) Proc Natl Acad Sci USA 99:3024
181. Dubey N, Letourneau PC, Tranquillo RT (1999) Exp Neurol 158:338
182. Dubey N, Letourneau PC, Tranquillo RT (2001) Biomaterials 22:1065
183. Schmidt CE, Leach JB (2003) Annu Rev Biomed Eng 5:293
184. Yang Y, De Laporte L, Rives CB, Jang JH, Lin WC, Shull KR, Shea LD (2005) J Control Release 104:433
185. Cao X, Shoichet MS (1999) Biomaterials 20:329
186. Rosner BI, Siegel RA, Grosberg A, Tranquillo RT (2003) Ann Biomed Eng 31:1383
187. Mahoney MJ, Saltzman WM (2001) Nat Biotechnol 19:934
188. Maysinger D, Krieglstein K, Filipovic-Grcic J, Sendtner M, Unsicker K, Richardson P (1996) Exp Neurol 138:177
189. Pittier R, Sauthier F, Hubbell JA, Hall H (2005) J Neurobiol 63:1
190. Woerly S, Pinet E, de Robertis L, Van Diep D, Bousmina M (2001) Biomaterials 22:1095
191. Lore AB, Hubbell JA, Bobb DS Jr, Ballinger ML, Loftin KL, Smith JW, Smyers ME, Garcia HD, Bittner GD (1999) J Neurosci 19:2442
192. Li Y, Wang J, Lee CG, Wang CY, Gao SJ, Tang GP, Ma YX, Yu H, Mao HQ, Leong KW, Wang S (2004) Gene Ther 11:109

Author Index Volumes 201–203

Subject Index

Printing: Krips bv, Meppel
Binding: Stürtz, Würzburg